50/-

80 6017⁰⁴¹

D1577473

14.

23.

25 OCT 1997

-1 MAY 200

4 MAY

Co

The Un

LF 75/5

BASIC GEOLOGY
FOR ENGINEERS

J. BUNDRED, M.A.(Cantab.), C.Eng., M.I.C.E., M.Inst.W., F.G.S.

*Professional Officer in the Department of
Highways and Transportation,
Greater London Council*

LONDON
BUTTERWORTHS

ENGLAND: BUTTERWORTH & CO. (PUBLISHERS) LTD.
                 LONDON: 88 Kingsway, W.C.2

AUSTRALIA: BUTTERWORTH & CO. (AUSTRALIA) LTD.
                 SYDNEY: 20 Loftus Street
                 MELBOURNE: 343 Little Collins Street
                 BRISBANE: 240 Queen Street

CANADA: BUTTERWORTH & CO. (CANADA) LTD.
                 TORONTO: 14 Curity Avenue, 374

NEW ZEALAND: BUTTERWORTH & CO. (NEW ZEALAND) LTD.
                 WELLINGTON: 49/51 Ballance Street
                 AUCKLAND: 35 High Street

SOUTH AFRICA: BUTTERWORTH & CO. (SOUTH AFRICA) LTD.
                 DURBAN: 33/35 Beach Grove

Suggested U.D.C. *No.* 55 : 624
Suggested additional No. 624 · 131 · 1
Standard book numbers 408 41950 4 case
                       408 41951 2 limp

Made and printed in Great Britain by
R. J. Acford Ltd., Industrial Estate, Chichester, Sussex

To

A.B., A.L. AND P.J.B.

PREFACE

The name *Geology* is a word derived from the Greek ('ge'—earth, 'logos'*—a scientific treatise or discourse) and therefore means a study of the *Science of the Earth.*

It embraces within its scope a knowledge of all the natural sciences, for in the study of the history, nature and structure of the crust of the Earth, contact with some facet of every known branch of the sciences is inevitable and it can be a stimulating and fascinating experience. To the present-day technologist or student, such study offers more than mere scientific or profitable reward in his chosen field, for it is bound up with fundamental philosophical questions of earth science and astronomy as to the origin and nature of the Earth, planets, stars and Universe, the place of our own planet therein, and the appearance, with his superior intellectual development and dominance over other species, of that recent evolutionary phenomenon, *'homo sapiens'.*

To fellow students interested in Earth phenomena, and in particular to the student technologist impelled with the future task of using his knowledge and practical acumen to the harnessing of Nature for the benefit of all mankind, this introductory book is modestly and respectfully offered.

No attempt has been made, however, except by listing suitable references to extend the scope of study beyond a detailed account of particular aspects immediately of interest to the engineer or builder, as there are many other excellent books on geology, to which references are given, that deal more adequately with the general aspects of Earth science. The omission of some topics of traditional Geology, such as a detailed stratigraphical study of the British Isles and Paleontology, has been inevitable within the limited space available, which has restricted the choice of material from a vast spectrum of possible topics. But it is hoped that the concentrated resumés provided for certain traditional aspects of geological teaching, supply enough background for this text and will enable the student to cover the groundwork satisfactorily.

The aim of the book is to describe the fundamental data and establish the general principles underlying the science of Engineering Geology, the main object being to assist students preparing for Associate Membership (via C.E.I. examinations) of the I.C.E. and I.Struct.E., and for Higher

---

* *'Logos'* is a common term used in ancient philosophy and religions to express the idea of a reasoning mind or power responsible for the ordered state of the world and the study/discussion of this order, as revealed in natural phenomena.

National Diplomas and Certificates in Building, Civil or Structural Engineering; for College Diplomas, University first degrees and/or Diplomas* in Technology. The whole of the contents should be well within the reach of a student approaching the subject for the first time.

The aim has been always to make the text sufficiently comprehensive within the bounds of condensation for a book of moderate length, which means that some facts had to be omitted and others treated lightly. For the obvious shortcomings and generalizations of an approach that reduces the traditional geological work to a minimum and for any unfortunate errors and omissions that may normally arise, the author apologizes.

Again, the readers' attention is drawn particularly to the references and suggestions for further reading at the end of each chapter, which it is hoped will prove valuable to those students whose interest is aroused to follow the subject of Geology a stage further. More generally, those students whose imagination is stimulated to study the fundamental question of mans' existence on this planet, should also find some topics, references and information of interest to them, which result the author trusts may in part be credited to this text.

* New C.N.A.A. Degrees.

London

J. BUNDRED

viii

ACKNOWLEDGEMENTS

It is impossible to acknowledge adequately the many and varied sources of information to which the author has had recourse during many years of study and practice, and so he makes general acknowledgement and extends his grateful thanks to the lecturers, authors, former colleagues, specialist firms and others whose advice has been a source of inspiration and assistance to him in the accumulation of the knowledge presented in this book.

He wishes especially to thank Miss E. Williams for undertaking the typing of the manuscript, Messrs. A. Latif and P. A. Green for reading it and making helpful suggestions regarding style and contents; also Mr. A. Latif for preparing various diagrams and assisting in the organization of the material, and would not wish to forget the generous advice and assistance of his publishers at all stages of production.

To the Council of the Institution of Civil Engineers for permission to quote from their examination papers, and to the Institute of Geological Sciences, the British Museum of Natural History, etc., for permission to reproduce the phototographs and references given, the author would also express his general thanks; acknowledgements of individual items are given where appropriate in the text.

There must inevitably be accidental errors and omissions in the text of this book and it is hoped that these will be reported by interested readers.

CONTENTS

xi

CONTENTS

CHAPTER 1

GEOLOGICAL STRUCTURE AND HISTORY OF THE EARTH

1.1. INTRODUCTORY REVIEW OF GEOLOGY

To carry out a detailed geological study of the earth entails first the listing of the materials of the earth's surface crust and interior, the latter mainly through *Geophysics* and *Seismology*; it would then require a cataloguing of the physical and chemical properties of all known rocks and minerals— including the relevant information regarding their crystalline formation from molten 'igneous magma' (*see* page 14), or by weathering and 'sedimentation' processes (*see* page 15), or through the 'metamorphism' of pre-existent rock. Each of these examples is a typical facet of *Geology* and a highly technical study in itself, furthered broadly by the sciences of *Petrology* and *Mineralogy*. Further, a comprehensive geological survey entails also the discovery of the distribution, physical structure and 'stratification' of the various types of surface rock layers through *Physical Geology* and *Stratigraphy*, and should include, in addition, information on the nature of oceans, seas and continents, together with their variations of topographic and climatic conditions throughout the world. Any complete study of the earth's geology should contain, as well, an attempt to interpret fossil records of the plant and animal life from previous ages, through the deductions of *Paleontology*.

In short, all the foregoing could lead to a detailed biography and structural analysis of the earth with rocks classified mainly by age and type as based on their stratification; then the application of this classified knowledge, through the conclusions of *Historical Geology*, to man's understanding of the present-day world.

The broad basis of geological studies and their points of contact with other related sciences such as geodesy, geography and cartography, physics, chemistry and biology, should thus be readily apparent, although our immediate purpose will be to describe the relevant scientific aspects of geology which relate to their practical uses in the various fields of modern industrial technologies. For example, in the exploitation of raw materials for various manufacturing processes by obtaining coal and iron ores, water or oil, brick clays, building sands and aggregates, chalky materials for cements and concretes, and so on—termed *Economic Geology*; in the siting and construction of dams, reservoirs, road or rail tunnels and other works of man—termed *Engineering Geology*. These multitudinous activities, through the partnership

1

DESCRIPTION OF EARTHQUAKE WAVES

Seismic L, P and S waves travel outwards from Epicentre and Focus as expanding spherical wave fronts - shown diagrammatically as rays normal to these fronts. The rays are bent by reflection at discontinuities or boundary surfaces such as those of CRUST-MANTLE and MANTLE-CORE. They are bent by refraction, through gradual changes in physical properties of the transmitting medium at depth - upwards at the Mantle, downwards at the Crust. P waves pass through both Core and Mantle, but S waves propagate through Mantle only.

Sedimentary layers generally overlie Sial and Sima. Thickness 1/2 mile average on land but up to 8 miles (12km) in former Geosynclinal Basins - over Sima ocean floor are layers up to 1 mile thick

Transmission of compressional ||||||||||||||||||| P (for push) and transverse ∿∿∿∿ S (for shear) shock waves through the solid elastic mantle. Both types of waves are recorded clearly at observing stations B, E, F and G (Table 1.2)

P, S and L waves received on seismographs

Broad angular ring of Shadow Zone forms a belt on Earth's surface surrounding a central P wave antipodal circular reception spot

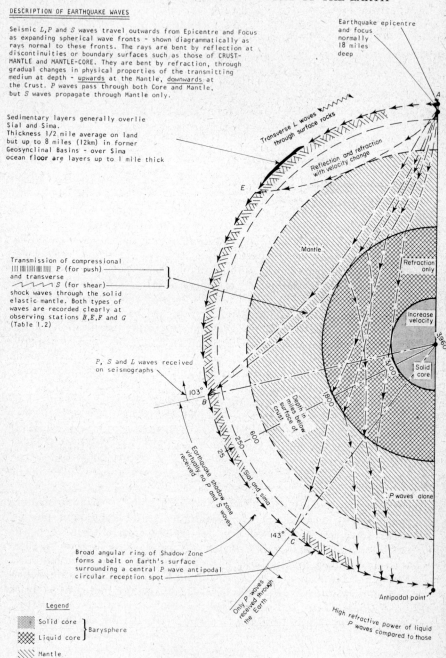

Legend

Solid core ⎫
Liquid core ⎬ Barysphere
Mantle

Figure 1.1. Structure of the Earth (cross-section through centre—radius 3,960 miles (6,400 km.)

1.1. INTRODUCTORY REVIEW OF GEOLOGY

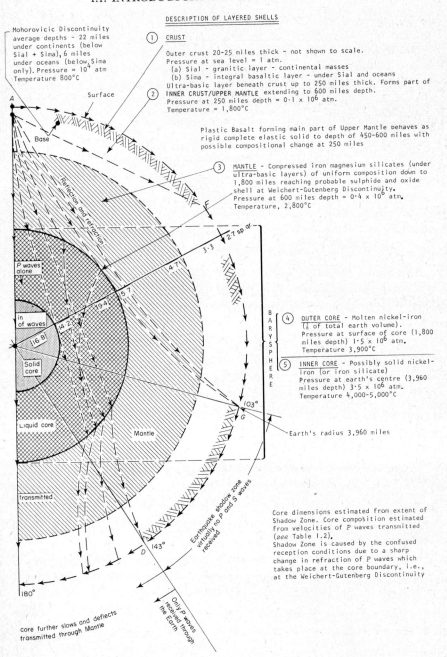

Mohorovicic Discontinuity
average depths - 22 miles
under continents (below
Sial + Sima), 6 miles
under oceans (below Sima
only). Pressure = 10^4 atm
Temperature 800°C

① CRUST

Outer crust 20-25 miles thick - not shown to scale.
Pressure at sea level = 1 atm.
  (a) Sial - granitic layer - continental masses
  (b) Sima - integral basaltic layer - under Sial and oceans
Ultra-basic layer beneath crust up to 250 miles thick. Forms part of
② INNER CRUST/UPPER MANTLE extending to 600 miles depth.
Pressure at 250 miles depth = 0.1×10^6 atm.
Temperature = 1,800°C

Plastic Basalt forming main part of Upper Mantle behaves as
rigid complete elastic solid to depth of 450-600 miles with
possible compositional change at 250 miles

③ MANTLE - Compressed iron magnesium silicates (under
ultra-basic layers) of uniform composition down to
1,800 miles reaching probable sulphide and oxide
shell at Weichert-Gutenberg Discontinuity.
Pressure at 600 miles depth = 0.4×10^6 atm.
Temperature, 2,800°C

④ OUTER CORE - Molten nickel-iron
(⅛ of total earth volume).
Pressure at surface of core (1,800
miles depth) 1.5×10^6 atm.
Temperature 3,900°C

⑤ INNER CORE - Possibly solid nickel-
iron (or iron silicate)
Pressure at earth's centre (3,960
miles depth) 3.5×10^6 atm.
Temperature 4,000-5,000°C

Earth's radius 3,960 miles

Core dimensions estimated from extent of
Shadow Zone. Core composition estimated
from velocities of P waves transmitted
(see Table 1.2).
Shadow Zone is caused by the confused
reception conditions due to a sharp
change in refraction of P waves which
takes place at the core boundary, i.e.,
at the Weichert-Gutenberg Discontinuity

Surface

A

Base

Refraction and refraction

P waves
alone

in
of waves

16.8

Solid
core

Liquid core

Mantle

transmitted

BARYSPHERE

2.7 sp.gr.

3.3

4.7

5.7

9.4

4.2

103°

G

143°

D

180°

Earthquake shadow zone
virtually no P and S waves
received

Only P waves
received through
the Earth

core further slows and deflects
transmitted through Mantle

3

of the academic scientist and practising technologist, form the basis of our present-day industrial civilization and its environment.

The student who prefers to commence his study from the estimated beginning of earth history and to read something about the earth's place in the solar system, as well as its presumed origin and development, is referred to Ryder (1949), Adams (1954), Hurley (1960), Lovell (1963), Gamow (1965) and Dunbar (1966), who give cosmological theories and data.

1.2. STRUCTURE OF THE EARTH

(see Figures 1.1, 1.2 and *1.3* and Table 1.1)

The surface area of the earth, 196,950,000 sq. miles, includes 57,510,000 sq. miles of land surface and 139,440,000 sq. miles of water area, so that only 29 per cent of the earth's crustal surface is visible as dry land at an average elevation of just half a mile above mean sea level, the major part of this surface being at an almost uniform depth of 3 miles (5 km) below mean sea level (M.S.L.).

The earth's 'visible' *Crust* consists largely of a light-coloured stony or solid granitic-type material of average sp.gr. 2·65 forming a very shallow sedimentary, igneous and metamorphic skin layer of rocks with varied compositions, and on average ½ mile in total thickness. This extreme outer skin, sometimes called the *Upper Lithosphere*, is composite with the main granitic crust (sp.gr. 2·7), or *Lithosphere*,* which itself grades into an inner shell layer of denser basaltic-type material (sp.gr. 3·0) comprising the earth's *Lower Lithosphere*.* Below these rigid crustal layers lies the much denser *Upper Mantle* zone, initially composed of an ultra-basic, volcanic-type material (sp.gr. 3·3) but grading down into the main *Mantle* shell (average sp.gr. 4·3) that extends inwards with ever-increasing density to a depth of 2,000 miles (3,000 km) until the earth's *Core* boundary is reached. (Mean radius of earth = 3,956 miles.)

The convenient and self-explanatory term *Sial* has been coined from the first letters Si and Al of the words silica and alumina, the predominant chemical constituents of the 'outer crustal layers' [silica = silicon dioxide = SiO_2, usually found in the crystalline form of the mineral quartz, or combined with alumina = Al_2O_3 as acidic alumino-silicate minerals], and is used to refer only to the continental land masses of the Lithosphere.

The similar term *Sima*, derived from the first letters of silica and magnesium, the principal chemical constituents of the dark volcanic 'basalt'-type lava layers (composed mainly of basic ferro-magnesian silicate minerals, such as olivine and augite), is often used as a short descriptive word for the *Lower*

---

* The introduction of many terms here and throughout Chap. 1 is deliberate, so that the student may learn them early in his studies and by constant repetition become familiar with their usage as the basic vocabulary of geology.

Table 1.1. A Summary of some Relevant Details of the Earth and its Crust

(a) Some Details of the Earth

Geodesy				
	miles²			
Superficial area of earth	196,950,000	Highest point of land surface:*	Mount Everest	29,028 ft above M.S.L.
Land surface (only 29 per cent of total area lies above M.S.L.)	57,510,000	Lowest point of land surface:*	Dead Sea, Palestine, in Northern end of Great Rift Valley. Shore level	1,286 ft below M.S.L.
Water surface (71 per cent of total area)	139,440,000			
	miles			
Polar diameter	7,901	Greatest ocean depth:*	Mariana's Trench in the Challenger Deep off the Philippines, Pacific Ocean	36,198 ft below M.S.L.
Equatorial diameter	7,927			
Mean diameter (12,750 km)	7,913			
Equatorial circumference	24,902			
	miles³			
Volume of the earth $(1\cdot08 \times 10^{17}\text{km}^3)$	26×10^{10}	Greatest mountain range*	Sub-oceanic mid-Atlantic Ridge with island peaks (e.g. Iceland, Azores, St. Helena) extending from N. Spitzbergen to Tristan Da Cunha	10,000 miles long × 500 miles wide
Mass of the earth $(5\cdot98 \times 10^{24}\text{kg})$	$5,887,613,230 \times 10^{12}$ tons			
Mean specific gravity of the earth	$5\cdot517$	Vulcanicity:†	The total number of active volcanoes is 485. The most violent recorded eruption was that of Krakatoa, Indonesia in 1883.	
Estimated age of the earth	$4,500–5,000 \times 10^6$ years			

* All represent mere wrinkles on the earth's crust (or that of any similar contracted sphere) with a total range of 12 miles—minute compared to a mean diameter of 7,913 miles.

† Earth Structure Globe (P/C MNL 532) and large map in London Geological Museum with exhibits, e.g., Diorama P/C MN 277, Vesuvius in eruption, 1872, should be inspected if possible. (*See* other references in Chaps. 1, 2 and 4, for earth structure and geography.)

Note: By comparison the mean diameter of the sun is 866,400 miles and sp.gr. 1·4. The earth makes a complete rotation on its axis in 23 hours 56 min. at mean equatorial speed > 1,000 mile/h and revolves in orbit around the sun in 365 days, 5 h, 48 min, 46 s (1 sidereal year) at an orbital velocity of 66,600 mile/h.

Lithosphere, which underlies both the continental land masses and oceans alike.

Thus, both Sial and Sima layers together comprise the earth's *Outer Crust* (i.e., the exterior rock shell down to the Upper Mantle/Inner Crust zone) with a total thickness of between 20–25 miles (30–40 km); under the mid-continental surfaces the Sial has an average thickness of 15 miles (25 km), but under the oceans where the Sial is largely absent, the Sima alone may be, in places, as little as only 3 miles (5 km) thick.

The *Mantle* shell itself surrounds a much denser, intensely hot and molten metallic interior known as the *Core* or *Barysphere* (from the Greek, *Baros* = weight); this, in turn at a depth of 3,000 miles (5,000 km), appears to be divided into an *inner solid central sphere* of 1,700 miles diameter (2,700 km) beneath its outer molten part. Geophysical observations confirm that the planet earth consists of these three major shells, *Crust*, *Mantle* and *Core*.

Table 1.1—*continued*

(b) *Principal Oceans and Some Seas*

Ocean/Sea	Area mile²	Average depth ft	Greatest depth ft
Pacific	63,986,000	14,040	36,198
Atlantic	31,530,000	12,800	30,143
Indian	28,350,000	13,000	32,968
Arctic	5,541,600	4,200	17,850
Mediterranean	1,145,000	4,500	14,435
North	221,000	180	1,998
Baltic	158,000	221	1,300

Summary: The ocean and sea areas amount to approximately two and a half times the land surface areas and these latter have undergone great changes since their formation (*see* Chaps. 2 and 5). The *major part* of the earth's surface is ocean bed at an average depth of approximately 5 km below M.S.L. and the *minor part*, the continental land masses, lie mainly in the Northern Hemisphere at an average height slightly above M.S.L. A very small percentage of the total surface area lies above, between or below these two levels as mountain chains, ocean deeps (± 10 km) and continental shelves (*Figure 2.20*); there is, however, a fundamental relationship between the average level of continental masses and that of the ocean beds (as shown by the *Theory of Isostasy*) and resulting from the force of the earth's gravity (*see* Chaps. 1, 2 and 4, References to earth structure, vulcanism, geophysics and geography).

The *Mantle* thus extends about half-way to the earth's centre and a scientific study of the transmission modes through this particular shell for seismic, *P* and *S*, shock waves emanating from any earthquake 'focus', with its corresponding surface 'epicentre' (*Figure 1.1*), reveals that the Mantle possesses all the properties of a layered plasto-elastic solid by transmitting both compression, *P*, and shear waves, *S*. The *Core*, however, behaves as a liquid since it does not transmit the shear (or transverse) waves thus produced but allows passage only to the compressional shock waves radiated (Table 1.2.); there is also a distinct discontinuity, called the 'Weichert–Gutenberg Discontinuity', occurring at the boundary of separation between liquid core and mantle.

Table 1.1 summarizes some relevant details of the earth.

	Chemical elements, percentage	Remarks	Mineral examples (primary)		Remarks (See Chap. 3 for full details)
The crustal layers are not always homogeneous and uniform in composition; local accumulations of a particular element or compound formed by natural processes often provide a useful source of commercial supplies*	Oxygen 47 Silicon 28	Together form a total of 75% of the elements included in all mineral compounds	Elements are found mainly in the form of chemical compounds as minerals	Quartz	Most abundant mineral compound of colourless or white silicon dioxide (SiO_2) forms crystalline 'matrix' in granite rocks in which the mineral crystals of pink and/or white Felspar and black and/or white mica are embedded and interspersed
				Felspars	—Orthoclase (pink) and plagioclase (white) are complex silicate compounds of silicon, oxygen and aluminium with either potassium, sodium or calcium
	Aluminium 8·00 Iron 4·50 Calcium 3·50 Sodium 2·75 Potassium 2·50 Magnesium 2·25	Altogether form a total of 24% of the elements* included in all mineral compounds		Mica	—is a complex ferro-magnesian silicate of aluminium, potassium, and hydrogen (white colour) or with magnesium and ferrous (iron) oxide (black colour)
	98·50+		Dark green to black in colour	Augite	—is a complex ferro-magnesian silicate of magnesium, ferrous (iron) oxide, calcium with aluminium
				Hornblende	—is a complex ferro-magnesian silicate of magnesium, ferrous (iron) oxide, calcium with aluminium and sodium
				Olivine	—is a complex ferro-magnesian silicate of magnesium and iron; it hydrates to form Serpentine

* The remaining 80 or so elements, including all those contained in living matter, comprise less than 1 per cent of the outer crustal layers. Many of the metallic elements—gold, silver, copper, zinc, lead form < 0.5 per cent of all crustal material, and can only be extracted from areas of concentration as in ore veins—such ores are economically workable, however, at very low local concentrations, for example:

Element	Average proportions in crustal rocks	Local concentration required for economic extraction
Gold	1:100,000,000 parts	If greater proportion than 1:100,000 parts of rock mass
Copper	1:10,000 parts	If greater proportion than 0·75 per cent of a rock mass or strata
Iron	compared to 1:20 parts	
Aluminium	1:12 parts	If greater proportion than 18–20 per cent of a rock mass or strata

7

1.3. FORMATION OF THE CRUST AND INTERIOR

The earth's 'outer crust', as mentioned in Section 1.2, is divided into two main layers, the uppermost being the Sial (or granitic layer) which for the continental blocks, rests, rather as a sheet of slag does on the surface of its molten liquid, in the underlying inner solid layer of the Sima (or basic basaltic material). The latter has a thickness varying between some 3–10 miles from sub-oceanic crust to mid-continental layer, although the 'Lithosphere (i.e. Sial + Sima together) may extend up to 25+ miles (40–50 km) in maximum thickness where a compensating granitic or basaltic rock mass (generally a smooth bulge many miles wide, with the Sial base perhaps 20 miles deep) lies beneath a continental mountain chain like the European Alps or Himalayas. This bulge phenomenon of Sial and Sima is similar in effect to the flotation of an iceberg in the sea and stems from a property of the crust known technically by the name of *Isostacy*,* a term coined by C. E. Dutton in 1899 and taken from the Greek meaning 'standing still'; 'Isostacy' expresses the fact of the equilibrium of different parts of the earth's outer crust above the Upper Mantle resulting from the force of gravity. The Sial blocks are immersed in the Sima with one-ninth of their total average thickness above M.S.L. and eight-ninths submerged, according to the Principle of Archimedes.

At the boundary surface between the Sial/Sima *Crust* and the *Mantle* a phase discontinuity in the earth's material occurs which is called the 'Mohorovic Discontinuity' (or 'Moho'† for short), after the name of the Croatian seismologist, A. Mohorovičic, who discovered it from seismic measurements in 1909. The 'Moho' boundary lies at different levels in various regions, but is generally found at 22 miles depth from M.S.L. under the land continents and at 6 miles depth under the oceans (*Figure 1.2*).

Below this discontinuity, which could indicate an actual change in the chemical composition of the underlying rock shell or merely, as is actually believed, a phase change of its physical state, the material of the Mantle is probably composed of heavy ultra-basic ferro-magnesian minerals; samples of these, like 'peridotite', 'picrite' and 'olivine' of sp.gr. 3·3+ (Table 1.1) are found on the floor of the Pacific Ocean, when freed from its covering of deposited sediments. Here also, in most places, the granitic Sial layer is either very thin or practically non-existent (*Figure 1.3*). The defunct 'Mohole Project'† of the U.S.A. was designed to discover the precise physical and chemical properties of the Upper Mantle material *in situ*, which properties, apart from the information provided by geophysical observations on the refraction and velocity changes of earthquake shock

* *See* Chap. 2, page 61, for a further description of Isostacy.

† Some details of the Mohole Project (and also JOIDES) are given in the References, pp. 26–27. This imaginative and interesting American plan to drill through the oceanic crust to below the Moho and confirm the supposed properties of the Mantle was well advanced until recently.

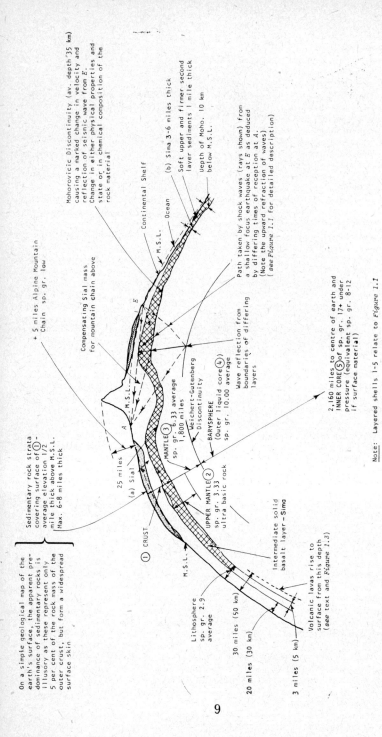

On a simple geological map of the earth's surface, the apparent predominance of sedimentary rocks is illusory as these represent only 5 per cent of the rock mass of the outer crust, but form a widespread surface skin

Sedimentary rock strata covering surface of ① - average elevation 1/2 mile thick above M.S.L. Max. 6-8 miles thick

+ 5 miles Alpine Mountain Chain sp. gr. low

Mohorovicic Discontinuity (av. depth 35 km) causing a marked change in velocity and reflection of seismic wave from E: change in either physical properties and state or in chemical composition of the rock material

Compensating Sial mass for mountain chain above

Continental Shelf

M.S.L.

Ocean   (b) Sima 3-6 miles thick

Soft upper and firmer second layer sediments 1 mile thick

Depth of Moho. 10 km below M.S.L.

Path taken by shock waves (rays shown) from a shallow focus earthquake at E as deduced by differing times of reception at A. (Note the upward refraction of waves) (see Figure 1.1 for detailed description)

M.S.L.

A

E

25 miles

(a) Sial

MANTLE ③ sp. gr. 6.33 average 1,800 miles

Weichert-Gutenberg Discontinuity

BARYSPHERE (Outer liquid core ④) sp. gr. 10.00 average

Wave reflection from boundaries of differing layers

① CRUST

UPPER MANTLE ② sp. gr. 3.33 ultra basic rock

Intermediate solid basalt layer—Sima

2,160 miles to centre of earth and INNER CORE ⑤ of sp. gr. 17+ under pressure (equivalent sp. gr. 8-12 if surface material)

Note: Layered shells 1-5 relate to Figure 1.1

M.S.L.

Lithosphere sp. gr. 2.9 average

30 miles (50 km)

20 miles (30 km)

3 miles (5 km)

Volcanic lavas rise to surface from this depth (see text and Figure 1.3)

CROSS-SECTION THROUGH CENTRE

NOT TO SCALE

Figure 1.2. Structure of the Earth—Enlarged details of Crust and Mantle

9

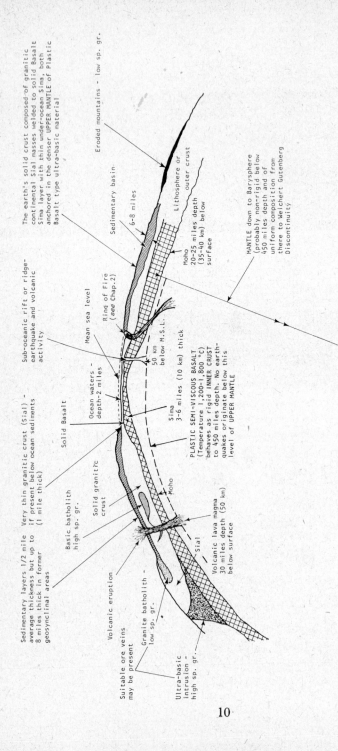

Figure 1.3. *Structure of the Earth—Enlarged detail—Rocks of the Lithosphere*

NOT TO SCALE

The earth's solid crust composed of granitic continental Sial masses welded to solid Basalt Sima layer with thin under-ocean Sima, both anchored in the denser UPPER MANTLE of Plastic Basalt type ultra-basic material

Eroded mountains - low sp. gr.

Lithosphere or outer crust

Sedimentary basin

6-8 miles

Moho
20-25 miles depth
(35-40 km) below surface

MANTLE down to Barysphere (probably non-rigid below 450 miles depth and of uniform composition from there to Weichert Gutenberg Discontinuity

CROSS-SECTION

Sub-oceanic rift or ridge-earthquake and volcanic activity

Ring of Fire
(*see* Chap.2)

Mean sea level

Ocean waters - depth-2 miles

Sima
3-6 miles (10 km) thick

50 km below M.S.L.

PLASTIC SEMI-VISCOUS BASALT (Temperature 1,200-1,800 °C) behaves as rigid INNER CRUST to 450 miles depth. No earthquakes originate below this level of UPPER MANTLE

Solid Basalt

Basic batholith high sp. gr.

Solid granitic crust

Moho

Sima

Volcanic lava magma 30 miles depth (50 km) below surface

Sedimentary layers 1/2 mile average thickness but up to 8 miles thick in former geosynclinal areas

Very thin granitic crust (Sial) - if present below ocean sediments (1 mile thick)

Volcanic eruption

Granite batholith - low sp. gr.

Ultra-basic intrusion - high sp. gr.

Suitable ore veins may be present

Note: All rocks are derived ultimately from the once fluid earth surface and primeval crust, no longer exposed to view. UPPER MANTLE belongs to the chemical category of 'magmas' as does the Sima and much of the Sial. Sima has a composition and density similar to that of the common black volcanic lava called Basalt and is thus referred to as the Basaltic layer of the Lithosphere

10

waves, still remain largely a matter for conjecture. (*Figures 1.2, 1.3* and Table 1.2.) Currently underway, project JOIDES (standing for the 'Joint Oceanographic *I*nstitutes *D*eep-*E*arth *S*ampling' programme of the U.S.A.) is to obtain cores from boreholes through the deep-ocean sediments of the Atlantic and Pacific Oceans in order to gather information on the oldest and undermost deposits 1,500 miles from land and at depths of 20,000 ft (6,000 m). The information to be collected should provide much new knowledge on the history and development of the earth's surface, possibly from the primeval crust stage. It may perhaps also enable samples to be obtained which represent a low-temperature version of the Upper Mantle material. (*See* Bascom (1961), Gaskell (1963), Wood and Blow (1968) and Tucker (1968).)

Table 1.2

Seismic velocities* (km/s)		Depth below surface (km)	Remarks
P waves	S waves		
8·2	4·35	33	'Moho'
11·42	6·36	1,000	
12·79	6·93	2,000	
13·64	7·30	2,900	'Weichert–Gutenburg Discontinuity'

Note: Of the three types of earthquake wave, the fastest:

    P or *Primary waves*, vibrate longitudinally in the direction in which they are travelling

    S or *Secondary waves*, vibrate transversely to the direction in which they are travelling

    L or *Long waves*†, the slowest, travel with transverse vibrations only through the crust

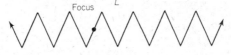

* Based on T. F. Gaskell (1963).
† '*L*' waves are named after 'Love' who first described them, and a second type of *L* wave is named after Lord Rayleigh.

    The earth's intensely hot central core is thought, at least in its outer shell, to be composed mainly of material which must be a molten iron–nickel or iron–silicate compound of average specific gravity between 10 and 12; this is a composition analogous to that of various metallic iron meteorites (normal sp.gr. 8) if their solid material were to be subjected, as a liquid mass, to the immense internal pressure and temperature existing in the *Barysphere*. Similarly, the deeper material of the *Mantle*, were it to be relieved of pressure and cooled solid, would be comparable to that of the stony meteorites* of normal sp.gr. 3·4 at ordinary temperatures. All the specific gravity values quoted have been translated to relate to material at a normal pressure, but in the earth's 'Inner core' the actual value of specific gravity is deduced as 17 for the solid centre portion and 14 for the fluid outer shell.

The layered structure of the earth shells is consistent with scientific conclusions on the earth's origin and life as a planet from the beginning of its primordial existence as whirling gaseous matter condensing to form layered shells, probably from two globular and immiscible fluids for the core and mantle materials, some 5 billion (10^9) years ago.

1.4. THE EARTH'S TEMPERATURE GRADIENT

Whatever hypothetical origin is assumed for 'planet Earth', all the available evidence points to the conclusion adopted in Section 1.3 that the *proto-Earth* (Gr. *protos* = first) soon became a very hot gaseous or perhaps fluid body, which cooled gradually as heat was radiated away from it. The young earth then slowly condensed until a layered solid crust formed upon it, crumpled and contorted, with emanated steam and gases forming a thick enveloping cloud and eventually, further cooling allowed the precipitation of rain and the formation of vast shallow seas on the solid crustal surface, the central core meanwhile remaining intensely hot.

As further evidence of the heat within the earth's interior, it is a well-known fact that the deeper one descends in a mine the higher the temperature becomes. The deepest mine descends about 2 miles or less than one-twentieth per cent of the earth's radius; yet at this depth the boiling point temperature of water has already been reached and this uncomfortable temperature increase is one limiting factor in the depth of mine workings. The temperature gradient is thus about a 1°C increase for each 100 ft of depth through the crust from the surface. In certain areas, e.g., the Yellowstone National Park, Wyoming, U.S.A., Iceland and at Rotorua, North Island, New Zealand, hot water geysers erupt through surface cracks owing to the effect of vapour pressure created in natural underground water reservoirs at a depth of approximately 1½ miles (2½ km) in the earth's outer crust.

Assuming that this temperature gradient remains constant, the Inner Crust/Upper Mantle temperature at a depth of 20–30 miles (50 km) must be such (1,200–1,800°C) that most known rocks would melt to become a

---

* Meteorites which fall on to the earth's surface from outer space are probably akin to the ring of asteroids which revolve around the sun between the orbits of the planets Mars and Jupiter, and are believed to be the fragments of a large planet originally pre-existent in this region of outer space. There are no meteorites yet found with a composition analogous to the superficial sedimentary rock layers formed by erosion processes on the earth's surface.

viscous fluid, were it not for the high internal pressure, 20,000 atm, prevailing there under the weight of overlying material (e.g., this pressure at a depth of 5 miles is of the order of 5–10 tonf/in²).

On the same basis the temperature of the outer core of the barysphere, at a pressure of 3.5×10^6 atm, is of the order of 4,000°C (7,200°F) and its material should thus exist in a completely molten state.

Not all the earth's heat, however, emanates from residual heat in its interior, transferred by conduction through the *Mantle* to the surface and radiated away from the outer crust, but much crustal heat is derived from the energy released during the transformation and spontaneous breakdown of radioactive mineral elements,* such as uranium and thorium. The variable and localized crustal distribution of these elements may greatly affect the *temperature gradient* in different mine workings, resulting, for example, in the South African gold mines having a gradient only about half that existing in British and European coal mines, there are, however, other localized effects such as the intrusion of very hot and molten volcanic matter into underground cracks or fissures near the earth's surface, as this magmatic igneous material (*see* footnote, page 14) rises from the lower Lithosphere or Upper Mantle zones (Section 1.6).

1.5. EARTH MOVEMENTS AND QUAKES—VOLCANOES†

Should the internal crustal pressure be lessened at any one point, then the underlying hot and semi-solid basic rock will liquefy and this can lead to a local instability in the earth's crust. Lava flowing from volcanoes originates from such a relief of pressure in comparatively shallow layers of the earth's crust when the plastic rock material becomes a viscous fluid *magma* (*Figure 1.3*), and may rise to the Sial surface as an 'igneous intrusion' or 'extrusion'; extrusive lava often forms a volcano, with its typical conical 'vent' through the build-up of this molten lava as it solidifies into rock either on land or under the sea.‡

The necessary temperature for this magmatic fluidity (1,200–1,800°C) is reached, as mentioned previously, at depths of 20–30 miles below the earth's surface and measured lava temperatures (1,200°C) inside the craters of existing volcanoes often correspond very well with the above hypothesis of magma formation.

---

* *See* Section 1.8. Actually, very little of the earth's original heat, stored at the time of its formation, has been and is being lost, because a major part of the energy required for this heat conduction and radiation is being produced by the breakdown of radioactive elements within the outer crust (and interior) of the earth, or in the near surface regions of the Sial.

† *See* Table 1.1 (*a*) under Vulcanicity.

‡ *See The Observer*, 17th Nov.–1 Dec. 1963, *Land of Fire Rises from the Ocean*, which describes the new volcanic island of Surtsey, south of Iceland and 3 miles from the Westman Islands at the northern end of the Mid-Atlantic Ridge. This eruption is probably connected with that which occurred in October 1961 on Tristan da Cunha, 10,000 miles away, at the southern end of the same Atlantic Ridge.

Volcanoes may be considered as analogous to safety valves of the Lithosphere, allowing the expansion of material from considerable depth in the Sima and upper layers of the Mantle, and thus relieving any local stress concentration in the crust. At depth in the Mantle, such semi-plastic viscous material under pressure, unlike the rigid brittle crust, can react as if perfectly elastic to the rapidly changing forces or 'shock waves' induced by earthquakes; this, despite the fact that the material may actually (if not permanently) be molten and would flow as a highly viscous fluid, albeit extremely slowly, given the time, relief of pressure and space to do so.

1.6. ROCKS OF THE LITHOSPHERE

Formation of High Temperature Igneous Rocks

It has been assumed that the original globular crust of the earth (possibly similar in composition to the Mantle material now existing) was formed by solidification of a primordial fluid mass, through condensation and cooling of its surface layers; we have explained also how volcanic action is continually erupting new molten magma upwards from the lower shell regions into the present 'outer crust'. Therefore, a lot of the superficial crustal material seen today was not necessarily formed as an age-old rock.

Molten magma which solidifies on the earth's surface is called *Extrusive Igneous* Rock* (forming the volcanic types), although much igneous rock never reaches the surface but cools deep underground and in fissures or faults of the overlying strata, being then known as *Intrusive Igneous Rock*. Thus igneous rocks are formed from parts of the earth's sub-crust, or Sima, which have melted and flowed upwards due to a local relief of pressure at some time, and only when subsequent denudation processes (Chap. 2, p. 33 et seq.) lead to exposure of major intrusions such as 'Batholiths' and 'Bosses' (*Figure 1.3*), are the coarse-grained slow-cooled crystalline rock types like granite found at the ground surface.

'Granitic intrusions' vary in their plan shape; they are commonly near circular at the ground surface, often several miles in diameter, and the surrounding rock strata are cut off sharply at the margin of the intrusion with an 'aureole' or halo of 'metamorphosed' (i.e., transformed) surface and country rock surrounding the granite mass; many granitic intrusions, in so far as they have been explored, appear to become wider as they are traced to greater and greater depth.

Rocks so formed are known as the Abyssal or Plutonic† types of 'major intrusions' while their more finely crystalline compatriots, the hypabyssal‡

---

* Igneous means literally *fire-formed*—in this case, *cooled from molten magma*, and magma denotes a *silicate melt* (or complex high temperature solution of silicates containing water/steam and various gases).

† The term Plutonic is derived from Pluto, the Greek god of the Underworld.

‡ Hypabyssal means not quite Abyssal, an adjective formed from the word Abyss—with an obvious connotation.

types of the minor intrusions, are formed from molten magma solidified near the surface in much smaller quantities, which were able to cool quickly in relatively narrow Dykes and Sills. Similarly, the *volcanic* rocks differ from the hypabyssal types in being yet more finely crystalline or even glass-like—as is the supercooled *obsidian* lava, for they have solidified very rapidly in contact with air or sea and often there was no time for any crystals to develop.

All these igneous rock types started as 'silicate melts', whose particular properties are governed mainly by their chemical composition and rate of cooling during solidification. The most characteristic property of magmas is their percentage of 'free' or combined silicon dioxide (free SiO_2 = pure silica); these acidic compounds vary in quantity between 35 per cent and 75 per cent of the total mineral compounds present in a 'melt' and determine the general colouring, density and silicate mineral content in aggregate, as compared to the 'basic' chemical composition, of any particular mass of igneous rock.*

A description of igneous rock types is given in Chap. 3; for students at present unacquainted with them, short descriptions of quartz, the two felspar types, mica and other important ferromagnesian minerals are given in Table 1.1 (*c*).

There are two other principal families of rocks:

Low-temperature Sedimentary Rocks

These rocks constitute most of the surface skin layers of the earth's 'visible crust' (i.e., as much of the outer Lithosphere as can be seen in quarries, mines, deep borings, etc.), to a maximum depth of 5 miles. They have been formed usually by the sedimentation of transported particles underwater in oceans, seas and lakes and in successive stratified layers according to age from the weathered-off fragments of pre-existing igneous, metamorphic, or other older rocks. As already stated in Section 1.2, these sedimentary rocks (or skin layers of the Outer Crust) generally cover their underlying Sial parent rocks to an average depth of about ½ mile, thus forming approximately 5 per cent of the total rock mass of the Outer Crust and by volume, equating to about 10 per cent of igneous rocks there present. Their formation and classification are described later in Chapters 2 and 3 and their detailed properties are given in Chapter 3.

---

* For details, *see* Mineralogy—Table 3.1, etc. The acid radical silica combines chemically with bases, i.e., oxides of sodium, potassium, aluminium, and also those of iron and magnesium, which react with the dissolved gases and water/steam in a 'magma' (or acidic rain water) to form 'salts' of silica (the 'acid' part of the silicates), the various original elements of the oxides being the 'basic' constituents of primary silicate minerals.

See Chap. 4 for further descriptions of the named and other related igneous phenomena of Section 1.6.

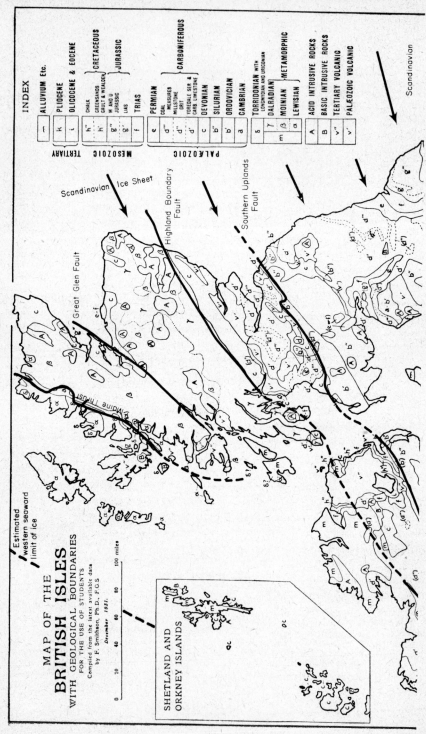

INDEX

	f	⌐	ALLUVIUM Etc.
TERTIARY	k		PLIOCENE
	i		OLIGOCENE & EOCENE
MESOZOIC	h"	CHALK	CRETACEOUS
	h'	GREENSANDS GAULT & WEALDEN	
	g"	M AND U JURASSIC	JURASSIC
	g'	LIAS	
	f		TRIAS
PALÆOZOIC	e	COAL	PERMIAN
	d"'	MEASURES	CARBONIFEROUS
	d"	MILLSTONE GRIT	
	d'	YOREDALE SER & CARB LIMESTONE	
	c		DEVONIAN
	b"		SILURIAN
	b'		ORDOVICIAN
	a		CAMBRIAN
	δ		TORRIDONIAN with LONGMYNDIAN AND URICONIAN
	γ	DALRADIAN	METAMORPHIC
	β	MOINIAN	
	α	LEWISIAN	
	A		ACID INTRUSIVE ROCKS
	B		BASIC INTRUSIVE ROCKS
	v"		TERTIARY VOLCANIC
	v'		PALÆOZOIC VOLCANIC

MAP OF THE
BRITISH ISLES
WITH GEOLOGICAL BOUNDARIES
FOR THE USE OF STUDENTS
Compiled from the latest available data
by F. Smithson, Ph.D., F.G.S.
December, 1931.

– – – Estimated western seaward limit of ice

0   20   40   60   80   100 miles

SHETLAND AND ORKNEY ISLANDS

Scandinavian Ice Sheet

Highland Boundary Fault

Southern Uplands Fault

Great Glen Fault

Moine Thrust

Scandinavian

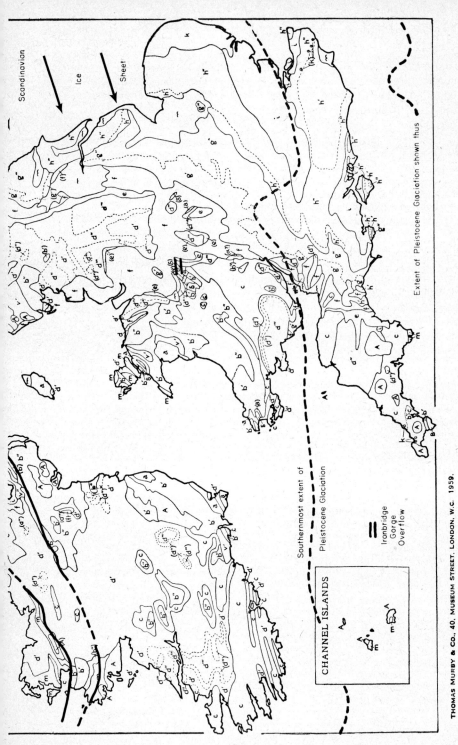

THOMAS MURBY & CO., 40, MUSEUM STREET, LONDON, W.C. 1959.

Figure 1.4. Geological Handmap of the British Isles
(From Platt, by courtesy of T. Murby and Co.)

Scandinavian

Ice

Sheet

Extent of Pleistocene Glaciation shown thus

Southernmost extent of

Pleistocene Glaciation

Ironbridge
Gorge
Overflow

CHANNEL ISLANDS

Metamorphic Rocks

Metamorphic rocks are derived from both igneous and sedimentary rocks which have been transformed and/or recrystallized (but often hardly altered in chemical composition) from their original state: this transformation occurs mainly by the action of heat and/or pressure supplied either by an 'igneous intrusion' to the rocks in its surrounding metamorphic aureole' (Chap. 4), or through the crustal rock stresses linked with an intensive earth movement.

1.7. BRIEF HISTORY OF THE DEVELOPMENT OF GEOLOGICAL SCIENCE

Some of the earliest attempts to explain the origin and nature of the Earth, like that by Aristotle (384–322 B.C.), author of *Physics and Metaphysical Ethics*, while reasonable enough from consideration of the evidence available to most ancient philosophers, were inevitably coloured by the mythological character of Greek and Roman thought and these ideas (developed later with Biblical and religious attachments) persisted for many centuries.

When, following on the Renaissance Period of learning in the Middle Ages, the French naturalist, Comte de Buffon (1707–1788) published his discourse of 1778 entitled, *Epoques de la Nature*,* this began at last a definitive trend toward the truly scientific approach which led finally to the development of the subject Geology in its modern form. Still earlier, in 1694, the pioneer geologist, John Woodward, had written a similar essay, *Towards the Natural History of the Earth*, and later again the Scotsman James Hutton (1726–97), sometime medical practitioner and agriculturalist, carefully studied the form of many rocks local to his home in Edinburgh, Scotland, together with their 'stratification'; he collected his factual observations scientifically, with an unbiased approach, in the age of a very prejudiced, conservative and religion-dominated society, especially with regard to the earth's creation and natural history. Hutton suggested an indefinite (or eternal) age for the earth, both past and future, when, in 1795, he published his massive treatise on *The Theory of the Earth*, a very revolutionary but scientific and scholarly work; indeed, Hutton proved the origin of igneous rocks and correctly interpreted the evidence of rock 'stratification' through the 'sedimentation' of waste material from weathered land surfaces. He realized also the significance of Unconformities (Chap. 4) and some of the reasons for the formation of folded rock strata into mountains.

William 'Strata' Smith (1769–1839) was a much travelled practising surveyor/engineer cum spare-time geologist, whose passionate hobby—the study of geology—enabled him to establish this subject as a worth-while science; by 1815, his observations enabled him to produce the first geological map of England and Wales, a truly incredible and wholly individual achievement. Canal Engineer Smith is often and aptly called 'The father of English

* This is only one of the 44 volumes of his brilliant and comprehensive *Histoire Naturelle*, or *Study of the Natural Sciences*.

Geology', both with regard to stratigraphy and as inventor of the geological map. He recognized, as useful scientific tools, the application of two basic laws which apply to all sedimentary 'rock strata', after his discovery that, to

(a)(i)   *(a)(ii)*   *(a)(iii)*

(b)

Figure 1.5. Typical fossils (a) (i) Fossil Crinoid (Sea Lily) from the Wenlock Limestone. Silurian—Older Paleozoic (ii) Leaf of a Coal Measures plant. Carboniferous—Newer Paleozoic (iii) An Ammonite from the Portland Beds. Jurassic—Mesozoic (b) Iguanodon—lower Cretaceous

((*a*) Crown copyright Geological Survey photograph. Reproduced by permission of the Controller, H.M. Stationery Office)
((b) By courtesy of Patrimoine, De L'institute Royal, Des Sciences Naturelles de Belgigue)

quote Smith's own words: 'the same strata were found always in the same order and contain the same peculiar fossils.' These two laws are summarized as follows:

1. For different rock strata, those that were originally lowest, are the older 'beds'.

2. Particular organic and fossil remains are characteristic of each rock layer or group of layers.

Fossil Records and the Age of the Earth—(relative time to absolute time)

The dating of sedimentary rock strata in their relative chronological order is achieved by careful study of their embedded fossil records, although clearly no such definitive record can exist for igneous and metamorphic rocks, due to their mode of formation.

Each period of earth history is exemplified by the fossilized remains found in the sedimentary rock material deposited during that period. The fossils thus discovered are mainly the embedded remains of ancient marine creatures (e.g., Ammonites), as these were easily buried in sea-deposited sediments; having a hard shelly structure, they became readily fossilized through the establishment of the earth's temperature gradient within the deposited material during its compaction into rock.

William Smith enunciated his famous principles, as stated above, mainly on the basis of his intimate knowledge of the *Mesozoic* (including Ammonite) fossils of S.W. England; this applied especially to the sea cliffs of W. Dorset, composed of shelly limestone beds and formed during the 'Lower Lias Period' (*Figure 1.6*), where the skeletal fossil remains of the large marine reptiles, Ichthyosaurus and Plesiosaurus have subsequently been found and some specimens extracted almost completely intact. Personal reference to Smith's work on the correlation of strata in S.W. Devon and, in 1808, the similar correlation by L. C. Cuvier of the 'cretaceous' strata from the Paris region of the Île de France with the chalk strata of S. England, is a most instructive exercise for any student. Recommended sources cited in Fenton (1945 and 1952) are given in the footnote, page 26.

Classification of Geological Eras, Periods and Rock Formations

All the foregoing leads on to the study by paleontologists of the fossil records in sedimentary rocks and their correlation for establishing the relative age of different strata, as illustrated in Table 1.3. The methods of Smith and Cuvier are today still the most accurate means of correlating strata in different geological regions and are therefore much used by the various oil, coal and other commercial companies' scientists in identifying and correlating the rock cores obtained by drilling, during their continual search for further sources of mineral wealth and fuels throughout the world.

To summarize—The early eighteenth–nineteenth century geologists had already discovered the means of classifying the order of the systems of sedimentary rocks into Eras, Periods and Sub-periods, whose 'relative ages' are calculable from their respective fossil records.

There remained for Smith (1815) and Charles Lyell (1839), 'the founder of modern geology', in the nineteenth century and later investigators only the establishment of an absolute time scale. Table 1.3 shows the final results of scientific and geological investigations for the stratigraphical succession of the sedimentary rocks as related to an absolute time base. This table has been deduced from the latest information on isotopic rock ages available to geologists through the methods of contemporary radioactive reearch. Table 1.3 also lists some relevant 'paleontological' data regarding the dominant life forms of each geological era or period.

Figure 1.6. Cliffs of Blue Lias, Lyme Regis, Dorset—rich in fossils
(Note the clearly defined sedimentary layers or typical bedded rock stratification of water-borne deposits)
(From C. P. Chatwin, 1936, by permission of the Controller, H.M.S.O.)

The names given to the various strata are mainly British in origin, since the classification of strata via the science of Stratigraphy, like modern Geology itself, originated and was predominantly developed in Great Britain.

1.8. RADIOACTIVITY AND ABSOLUTE GEOLOGICAL TIME

1. The following conclusions, due initially to the work of Lord Rutherford in 1905, Lord Rayleigh and others, are derived from a study of the breakdown or 'decay' of radioactive mineral elements present in rocks.

21

Table 1.3. Geological Systems and Stratigraphy

Era or Group (Figures = Time span × 10⁶ years) [ ] = Total Isotopic age	System or period	Sub-periods— Names of principal sedimentary rock strata and regions
QUATERNARY (meaning Fourth in order) OR Post-tertiary	Holocene	Soil, sand, Alluvium
	Pleistocene (a) [1] (Ice Age)	Glacial drift, Boulder clay, gravels, sands
CAINOZOIC from Greek 'Kainos' = recent. CAINOZOIC ('recent life') or (TERTIARY) 70	Pliocene (b) 11	Cromer series / East Anglian crags
	Miocene (c) [25] Alpine folding	Sands, clays, shelly deposits / No British examples
	Oligocene (d) 45	Isle of Wight limestones and marls
	Eocene (e) [70]	Bagshot Sands, London Clay, Thanet Sand
MESOZOIC = Middle life MESOZOIC (Secondary) 155	Cretaceous 65 [135]	Chalk / Greensand / Gault / Wealden } clays
	Jurassic 45 [180]	Purbeck / Portland } limestones / Kimmeridgian clays / Corallian limestones / Oxford clays / Oolites, Bathstone / Ironstone / Lias
	New Red Sandstone — Triassic 45 [225]	Rhaetic shales, Marls / Keuper Marls / Conglomerates / Bunter sandstones
	New Red Sandstone — Permian 45 [270] Hercynian Movements	Magnesian Lst. / Penrith Sdst.
EOZOIC = Ancient life PALEOZOIC 375 (Primary) — Newer or Upper 175	Carboniferous [355] 85	Upper { Coal Measures, Millstone grit and Sandstones; Lower { Carboniferous limestone
	Devonian (Old red sandstone) 50 [400] Caledonian Movements	Marine and terrestrial deposits *older than* Coal Measures of Carboniferous
Older or Lower 200	Silurian 40 [440]	Downton Beds / Ludlow Beds / Wenlock Beds / Llandovery Beds — limestones
	Ordovician 60 [500]	Bala Series / Llandeilo / Arenig } Grits and slates
	Cambrian 100 [600]	Tremadoc { Slates / Quartzites; Lingula Flags / Menevian Beds / Harlech Beds
CAMBRIA (Wales (Roman name) PRE-CAMBRIAN or (archaean) 1500	Torridonian / Charnian Revolution / Uriconian	Sedimentary { Grits, Sandstones, Quartzites
	Lewisian / Moineian / Dalradian Series	Igneous; Metamorphic

Laurentian Revolution   [2100]
Earth's primeval crust forming from liquid, after condensation of primordial inter-stellar gas-globe. [4500+]

Notes: 1. The names of the sub-periods are nearly all taken from the region of Great Britain where these rocks and their fossils were studied.
    2. *See* Chap. 3 for details of sedimentary rocks.
  * The so-called *Age of Reptiles* is better termed the *Age of Ammonites* as far as Great Britain is concerned. These molluscs are to be found plentifully embedded in the Mesozoic strata (e.g., at Lyme Regis and Charmouth, West Dorset, in the shale and limestone beds of the lower lias) but unfortunately such vertebrates as those of Ichthyosaurian origin are to be found only occasionally in certain special localities; other reptilian remains are also very rare and sparsely distributed. *See* Arkell (1947).

Table 1.3. Geological Systems and Stratigraphy

Dominant life form —related to fossil record	*Evolutionary movements*	*Origin of names for rock strata*
Modern Man 'Homo-Sapiens' 'Cro-Magnon', MAN 'Neanderthal'—'Swanscombe'	'Age of Man' Human and simian stocks separating. Anthropoids developing. Rise of mammals ↑	'Cene', from 'Kainos' as termination of words (*a*) Most of recent (fossils) (*b*) More .... (*c*) Less .... (*d*) Few .... (*e*) Dawn of recent life forms
'Pithecanthropus'—first true 'homo' 'Australopithecus—man ape Primates 'Proconsol'—primitive ape. Elephants, water birds, sharks and bony fish Grass-eating mammals Mollusc-shell fish, Modern plants and birds		
On land Dinosaurs, Pterosaurs Giant reptiles Reptiles and ammonites *In sea* Ichthyosaurus and aquatic reptiles Conifers and ferns	Extinction of Giant reptiles Appearance of flowering plants and deciduous trees. *Iguanodon* First birds appear *Age of Reptiles** Pterodactylus and Flying fish Land reptiles evolving Development of carnivorous fish-life reptiles and 'Ichthyosaurus'	Cretaceous—chalky Clays, are named after Weald District, England Purbeck, etc., after English localities and towns Jurassic after Jura mountains, Europe Oolites, after rock texture Corallian after Coral Triassic = threefold *New* Red Sandstone, *later* than Coal Measures
End of dominance of marine creatures	Many ancient forms of marine life become extinct, animal and life plant on land evolving	'Permian'—Russian province of Perm. Penrith—from Westmorland town, England
Amphibia, corals and 'Sea lilies'	First reptiles and insects Giant evergreen trees	'Carboniferous' from Coal measures
Abundant fossils from the 'Age Fish'	Rise of amphibia—e.g., 'Coelacanth'†—link with	After Devon County, England
(evolve into) →Ancestors of all modern types developing	Amphibia—on to Cretaceous and present day fish	
←Trilobites Graptolites Brachiopods	First land plants (leafless)	Shropshire and Welsh border localities 'Silures'—S. Wales ancient celtic tribe
First marine forms—soft bodied	First fish Rise of corals and molluscs	Welsh towns and localities 'Ordovices'—N. Wales ancient celtic tribe N. Wales locality Fossil remains, 'Lingula' N. Wales N. Wales town
'Proterozoic' = 'first life' shown by Doubtful fossils Fossil algae	'Azoic' = No fossils 'Eozoic' = a few soft-bodied life forms	After Loch Torridon, Wester Ross, Scotland Uriconium, Roman town near Shrewsbury, England Isle of Lewis, Hebrides

Rock beds of some ocean basins today are probably representative of the primeval crust, e.g., 'Pacific Moon Scar' depths like Marianas Trench (Table 1.1) where ancient basaltic and peridotitic lays immediately under the present day ocean sediments.

† *See* Smith (1956) on the modern coelacanth and its link with the Devonian Period. A specimen may be viewed at the British Museum (Natural History Section).

See also the famous work of Miller and Agassiz on Primitive Fish (cited by Fenton, 1945) . For fossils, see especially the publications of the British Museum, Natural History Section, e.g., *British Cainozoic Fossils,* 1959 and *British Mesozoic Fossils,* 1961 et seq. The bibliographies at the end of these publications give many further valuable references, e.g., Owen, R. (1851–64). *Reptilia of the Cretaceous Formations*; (1863–4). *Reptilia of the Wealdon and Purbeck Formations,* and (1874–9) *Reptilia of the Mesozoic Formations.*

See Davies (1947) for Paleontology and Harland (1967) *The Fossil Record.*

Many rocks contain minute quantities of radioactive minerals; the unstable isotopes of the elements uranium and thorium, for example, change spontaneously and slowly atom by atom, but at a known regular rate, into the stable helium and lead isotopes, liberating heat energy during this process. The rate of transformation of such radioactive elements is independent of the pressure, temperature, physical and chemical conditions of their surroundings and remains remarkably constant with time; thus these elements form a most dependable time-measuring instrument (or 'geological clock'), even on the scale required for that of geological eras and periods, e.g., Uranium 238 transforms into Lead 206 during a 'half-life' period of 4,500 million years. Thorium 232 transforms into Lead 208 during a 'half-life' period of 15,000 million years. (Note: the lead isotope so formed is not the same as ordinary lead in common use.)

In radioactive dating, the proportion of lead to uranium isotopes (or other such radioactive element) in a rock mineral, known as the *Uranium–lead ratio*, gives an estimate of the true length of time since the formation of the solid rock mass containing this particular mineral. However, to use this information satisfactorily, such radioactive minerals must be studied in parts of successive geological strata which have been unaffected as yet by weathering processes, and the mineral (e.g., like quartz) must itself be chemically stable in the presence of the various denudation agents.

From transformation values, such as those given above for the 'half-life' of uranium and of thorium, and using rock minerals assumed to be originally from the commencement of their radioactive breakdown, an integral part of and the same age as the solid rock in the strata containing them, it has been calculated that:

The Tertiary System has a minimum age of 30–60 million years
The Carboniferous System has a minimum age of 300–350 million years
The Devonian System has a minimum age of 360–400 million years
The Pre-Cambrian System has a minimum age of 600–2,100 million years

and this latter accounts for almost three-quarters of the whole geological time scale. But it should be noted that the earth must be older than the oldest sedimentary rocks as found today in the 'Lower Pre-Cambrian Series' of Scandinavia, with a probable age of $1,580 \times 10^6$ years (Table 1.3); these could only have been deposited after the condensation of water on the earth's surface and the long-term action of weathering processes on such pre-existing crustal rock as was then already formed—possibly equivalent to the present under-ocean Sima and Upper Mantle materials. As an indication of the extremely low concentration of radioactive mineral elements in ordinary rock, the following figures are given:

One ton of granite contains 9 g of uranium and 20 g of thorium,
One ton of basalt contains 3·5 g of uranium and 7·7 g of thorium,

The radioactive energy content however, is still enormous, e.g., the heat liberated by local concentrations of these minerals within the earth's crust may be a cause of the liquefaction of the lower Sima rock layers, thus producing igneous magma eruptions and crustal earthquake movements.* The age of sedimentary strata is generally determined by the study of radioactive minerals present in such igneous intrusions as they may contain and deduced from geological evidence to be of the same age as themselves. John Joly (1909), was the first to recognize the effect of radioactivity on the earth's heat content and as the energy supply for the formation of magmas and indeed mountains.

2. In 1910, *Professor N. H. Russell*, the famous physicist, obtained a maximum possible value for 'earth age' of 3,400 million years based on the quantity of 'transformed' isotopic lead now present in the earth's crustal layers and assuming that all of it was produced from the radioactive breakdown of the total uranium and thorium quantity in the original crustal minerals and rocks. It is thought by geophysicists and scientists today that radioactive material is definitely limited to the upper 20 miles of the earth's 'outer crust', i.e., the Sial and sedimentary skin layers of the Lithosphere. However, the earth may not be quite as old as the above figures suggest, since some lead present in the outer crust must have been formed as a 'primary mineral' and not as the end product of radioactivity.

3. During recent years a new method of measuring geological age has developed from the radioactive breakdown of the unstable element rubidium 87, which is transformed into stable strontium 87 in a half-life period of 60 billion (10^9) years. This method has shown that previous estimates of rock ages have always been too low; in some cases the previously accepted values have now been more than doubled and the method gives an estimated time of 5 billion (10^9) years back to the formation of the earth's original primordial crust. Starting from recently formed rocks, it would appear that the minimal age for the formation of the earth's primeval crust out of material which is largely the same as much still evident today, must be at least 2·8–3·5 billion (10^9) years for the commencement of geological history.

The reason for this revaluation is that the uranium or thorium–lead radioactive processes produce a 'radium gas emanation', radon–helium, which leads to the diffusion of some material away from its source and hence a loss in the amount of lead remaining behind as evidence in the vicinity of the uranium, etc., source rocks. Estimates of 'earth age' based on radioactivity of these uranium–thorium elements are necessarily minimal values, whereas the rubidium–strontium reaction, which proceeds without any such mass loss to affect the quantity of its end product, strontium, is a more definite 'hour-glass' running down through the millions of years of geological time. Hence the amount of strontium now left deposited near rubidium should give a reasonably accurate value for the date at which its igneous

* *See* footnote, page 13.

'parent' rock solidified in the outer crust or was intruded into a particular overlying sedimentary rock stratum.

Summarizing therefore, it would appear that the age of the earth's oldest sedimentary crust of rocks is approximately between 2,000 and 2,500 million years. Appropriate values have thus been inserted in Table 1.3 for the various successive geological eras and periods in the light of the above conclusions. Of the possible 5 billion (10^9) years of earth history, only the last 600 million years can be traced with reasonable accuracy in relation to geology and fossil records.

REFERENCES AND BIBLIOGRAPHY

Adams, F. D. (1954). *The Birth and Development of the Geological Sciences.* New York; Dover Publications

Arkell, W. J. (1947). The Geology of the Country around Weymouth, Swanage, Corfe and Lulworth. *Mem. geol. Surv. U.K.* London; H.M.S.O.

Bailey, E. R. (1967). *James Hutton—The Founder of Modern Geology.* London; Elsevier

Bascom, W. (1961). *A Hole in the Bottom of the Sea: the Story of the Mohole Project.* London; Weidenfeld and Nicolson

Beiser, A. and Editors Life Magazine (1964). *The Earth.* Aylesbury; Nelson

Bibliography of Seismology (1967). Serials. New Series, Vol. 1, No. 1. London; Geological Survey

Bowen, W. L. (1963). *The Evolution of the Igneous Rocks.* London; Constable

British Museum (Natural History) (1959). *British Caenozoic Fossils.* London; H.M.S.O.

British Museum (Natural History) (1962). *British Mesozoic Fossils.* London; H.M.S.O.

Bullen, K. E. (1954). *Seismology.* London; Methuen

Challinor, J. (1967). *Dictionary of Geology,* 3rd Edn. Cardiff; University of Wales

Chatwin, C. P. (1936). *The Hampshire Basin and Adjoining Areas.* London; Institute of Geological Sciences; H.M.S.O.

Davies, A. Morley (1947). *An Introduction to Paleontology.* London; T. Murby

Davison, J. (1954). *Great Earthquakes.* London; T. Murby

Dunbar, C. O. (1966). *The Earth.* London; Weidenfeld and Nicolson

*Fenton, C. L. and Fenton, M. H. (1945), 2nd edn. 1952. *Giants of Geology.* New York; Dover Publications

Gamow, C. (1960). *Biography of Earth.* 2nd edn. 1962. London; Macmillan

Gamow, C. (1965). *A Planet called Earth.* London; Macmillan

Gaskell, T. F. (1963). 'Boring Deep into the Earth'. *Proc. Instn. civ. Engrs.* **26,** 369

Gaskell, T. F. (Editor) (1967). *The Earth's Mantle.* London and New York; Academic Press

Geikie, A. (1905). *The Founders of Geology.* 2nd edn. New York and London; Macmillan

Graham, W. (1962). *Principles of Stratigraphy.* London; Constable

* The following references relating to page 20 and all other 'Giants of Geology' are cited by Fenton (1945) (and Adams, 1954) who include detailed accounts of these references in their works.

Hutton, J. (1788). 'Theory of the Earth'. *Trans. R. Soc. Edinb.,* **1** (Facsimile reprint 1959. New York; Hafner Publ. Co.)

Playfair, J. (1802). *Illustrations of the Huttonian Theory.* Edinburgh; Cadell and Davies and W. Creech. (Facsimile reprint 1956; Urbane. Ill.; Univ. of Illinois Press)

Smith, W. (1815). *A Delineation of the Strata of England and Wales with Parts of Scotland, etc.* London; John Cary. *See also* Eyles, J. M. (1969). *T. Soc. Biblphy nat. hist* 5 (2) 87–109, London, B.M.N.H. for all W. Smith's published works

REFERENCES AND BIBLIOGRAPHY

Harland, W. B. (1964). *The Earth*. London; Vista Books Ltd.
Harland, W. B., *et al.* (Editors) (1964). *The Phanerozoic Time-scale*. (1967). *The Fossil Record*. London; Geological Society
Howell, B. F. jnr. (1959). *Introduction to Geophysics*. New York; McGraw-Hill
Hurley, P. M. (1960). *How Old is the Earth*. London; Heinemann
Joly, J. (1909). *Radioactivity in Geology*. London; Constable
Leet, L. D. and Judson, S. (1965). *Physical Geology*. 3rd edn. Prentice Hall
Lovell, B. and J. (1963). *Discovering the Universe*. London; E. Benn Ltd.
Lyell, C. (1839). *Elements of Geology (and other works)*. London; Murray
Milne, D. A. (1954). *Earthquakes*. London; T. Murby
Poldevaart, A. (1955). *The Crust of the Earth. Special Paper 62*. Geological Society of America
Regional Catalogue of Earthquakes (1964). Serial of Geological Society **1**, No. 1. London (1967 onwards)
Ryder, T. A. (1949). *Mother Earth*. London; Hutchinsons
Sabine, P. A., Watson, J. *et al.* (1968). 'Isotopic Age-determinations of Rocks of the British Isles, 1955–64'. *Q. Jl. geol. Soc. Lond.* **123,** No. 492
Smith, J. L. B. (1956). Old Four Legs (Coelacanth). London; Longmans Green
Stamp, L. D. (1957). *An Introduction to Stratigraphy (British Isles)*. 3rd edn. London; Allen and Unwin
Tucker, A. (1968). 'History from the Ocean Bed'. *The Guardian*. 9 April
Wood, G. V. and Blow, W. H. (1968). 'An Introduction to JOIDES'. *Geol. Soc. Circ. No. 146*, April; (1969). *Proc. geol. Soc. Lond.* No. 1651, March, p. 213

CHAPTER 2

PHYSICAL GEOLOGY

2.1. INTRODUCTION

We have already listed many of the sedimentary rock types in Table 1.3; the brief classification of Table 3.7 shows typical examples.

By 'rock' the geologist means 'any aggregate of mineral particles' which may be material, ranging for example, from a loose uncemented gravel–sand or soft plastic clay to a hard solid igneous granite, and metamorphosed quartzite, or a strongly cemented rigid sedimentary deposit like sandstone. A 'mineral' is a single substance with a fixed chemical composition and, for the engineer, a practical definition of unconsolidated sediment or 'soil' is, 'matter loose enough to be dug with a spade'.

We must now consider the natural processes which bring about the breakdown or recrystallization of the igneous rocks and primary minerals, and convert their fragments on recrystallization into the other two broad families, sedimentary and metamorphic rocks. We may then proceed to consider these groupings in more detail.

2.2. SEDIMENTARY ROCKS

Sedimentary rocks which cover the larger part (i.e., three-quarters) of the earth's visible land surfaces, are always stratified or layered since their material has been deposited over wide areas, usually under water, in regular sheets one above another; their order and age of succession is used (Table 1.3) as the basis for a description of earth history. Three main sedimentary rock groups are generally recognized: *fragmental or clastic, chemical* and *organic.** Since all rocks suffer either disintegration or decomposition by weathering processes, at or near the earth's surface if exposed to atmospheric agents, this grouping forms a natural delineation of their resultant products.

'Disintegration' is the result of the mechanical processes of weathering, e.g., those due to rain, frost, wind; 'decomposition' is the result of chemical attack on the mineral constituents of rock.

Thus *fragmental rocks* consist of fragments of igneous and/or other pre-existing rocks, together with particles of some of the primary minerals originally formed in igneous magma (such as the predominant and un-alterable constituent of nearly all surface rocks, hard durable crystalline quartz), and they also contain secondary minerals which are the resulting

* These groups are then further subdivided for more convenience into those listed in Table 3.7.

compounds produced by chemical weathering (e.g., the clays, formed from such primary silicate minerals as felspar). Due to their wide and varied range of composition, fragmental rocks are commonly classified on the basis of grain size when forming uncompacted or loose sediments, as are, for example, gravels, sands, clays and mud.

Chemical sediments, on the other hand, are the result of precipitation of single substances from solution, and *organic deposits* consist predominantly of the remains of plant and animal life.

Typical examples of each type, consolidated and unconsolidated, are given in Table 2.1.

Table 2.1. Types of Sediment

Fragmental or mechanically formed		Chemically formed from solution		Organically formed	
Consolidated	*Unconsolidated*	*Consolidated*	*Unconsolidated*	*Consolidated*	*Unconsolidated*
Conglomerates	Boulders, Cobbles, Gravels, Coarse sand	Limestone, $CaCO_3$ Oolitic limestone (honeycomb type grains)	Chalk, Oolite Grains	Brown and bituminous coal, Anthracite	Peat, lignite decomposed plant matter
Sandstones	Coarse, Medium, Fine (sands)	Ironstone	Clay impregnated with Siderite (Carbonate of iron)	Crinoidal Limestone	Crinoid remains of 'sea-lily' or star fish
Siltstones, Shales, Mudstones	Silts, Muds and clays	Gypsum rock, Rock salt, Potash and soda salts, Dolomite (magnesian limestone)	Salt lake precipitations, Chalk interacting with magnesium salts	Shelly or coral limestone, Phosphorite Diatomite	Shell rubble and sand, Coral reef, Guano Diatomaceous earth

2.3. WEATHERING AGENTS

There are many causes of the wearing down of land masses, called 'denudation' and the subsequent building-up of new land areas by deposition of the resulting sediments.* We may classify weathering agents of denudation under three main headings, namely 'Water', 'Air' and 'Living matter', leading either to disintegration of rock particles through the action of rain, etc. (predominantly mechanical weathering), or to their decomposition by the acids in rain or river water (predominately chemical weathering). These agents would act on a land surface only to a limited extent, were it not for the transportation of their products through the action of gravity, rivers, wind and ice. The accumulation of debris or rock waste, where formed, leads to stagnation of the weathering processes on that area; but the constant removal of this waste material exposes a land surface to further attack, so that the weathering process becomes continuous. It should be realized that the various agents of weathering and transportation, although here described separately for convenience, are often inseparable in their order of importance and effect upon the form of a land mass.

* This deposition leads ultimately to regional earth movements and the elevation of the sea floor in the deposition zone. At the present time the continental masses are being reduced by an average amount of 0·02 mm annually. (*See also* pages 41, 67.)

Weathering by Water

(a) *Rain*—including chemical effects. Rain has its main effect on virgin rock by the scouring action of rivulets and dripping water, washing loose particles down to lower levels. The effects are most noticeable in tropical climates under very heavy rainfall, particularly in regions of poor vegetation. 'Gullies' of considerable size are often formed; soft rocks are removed and 'earth pillars' created where the earth, as a mixture of pebbles or boulders with fine sand or clay, is left capped by stone and the softer surrounding earth has been washed away (*Figure 2.1*). This effect is apparent in miniature on many earth slopes after rain, but outstanding examples of tall pillars are to be found in Scotland, the Austrian Tyrol and the Bad Lands (Nebraska) of the U.S.A.

Vegetation protects earth from the effects of rain. On the other hand, chemical effects produced by carbon dioxide (CO_2) (which is present in the average proportion of 3 parts per 10,000 by volume in the earth's atmosphere)

Figure 2.1. The Dens, 1 mile north of Fortrose, Scotland. (Glacial sands overlying boulder-clay filling old valley and now being eroded by the present Rosemarkie Burn and by the direct action of rain, giving rise to arêtes and earth pillars, which remain standing because the weight of capping rock has consolidated the pillar material more than the surrounding earth. The protective remains of soil and vegetation on the pillars was once continuous with that seen in the background, and serves as an index to the amount of material removed by denudation)

and sulphur dioxide (SO_2), both forming a weak acid solution in rain-water, cause attack on rock and removal of the resulting compounds such as bicarbonates in solution.   These weak acids attack limestone particularly and other carbonate rocks or rock minerals, producing 'swallow holes', 'clints' and 'grikes', and 'caves'* (e.g., the Cheddar caves and Wookey Hole in Somerset, Jenolan Caves, N.S.W., Australia) with the formation of 'stalactites' and 'stalagmites' by the slow deposition of minute globules of calcium carbonate left behind during evaporation of the falling water droplets. The same weathering agents are often the cause of comparatively rapid damage to many building stones, especially in the acidulated atmospheres of industrial towns.

(*b*) *Frost*—the effects of frost evident in the jagged serrated appearance of mountain peaks or ridges is the result of water entering cracks in rock surfaces, or being absorbed into their porous outer layers and later freezing to form a 10 per cent greater volume of ice. Chips spall off the rock surface or the surface grains are crushed and made more liable to decomposition; the surface affected then becomes friable and disintegrates under the action of the large pressures (about 2,000 lbf/in²) exerted by the expanded volume of ice. Broken-off fragments of crumbled rock later fall down to form 'screes' which are gradually weathered away by rain and streams of water.   Fine particles form a dust which is carried away by either wind or rain so exposing the underlying surface to further attack and, under the action of gravity, the fragmented rocks are themselves further agents of erosion.   Frost can cause soil to 'creep' down slopes, as in 'solifluction' movements, and also causes a *heave* of ground, especially in silty types of soil where road or runway slabs may be seriously affected by cracking (*Figure 2.2*).

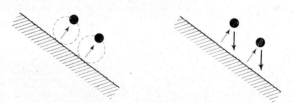

Figure 2.2 Frost heave
(Soil grains are lifted by 'frost/ice' at ground surface and fall back under gravity lower down a slope when thawing occurs)

Weathering in Air

(*c*) *Heat and cold causing Insolation*–Insolation is the weakening of a rock in course of time by the continual thermal expansion and contraction of the rock's various interlocked mineral constituents when subjected to repeated extremes of temperature.   In tropical and mountain regions this occurs

* *See* Section 2.8—Practical work No. 1;   London Geol. Mus. P/C MNL 515, A6038, MN 17081, CT19 and 53, etc., and the Wealden District for 'Swallow holes'.

especially between day and night, when the difference between maximum and minimum temperatures may be as much as 40°C, as at Timbuctoo, N. Africa (41°C) and Arizona, U.S.A. (34°C). The differential expansion and contraction of the mineral crystals sets up internal stresses which cause ultimate failure and disintegration at the rock surface. For example, quartz has a coefficient of cubic expansion of 0·000065 (65×10^{-6}) per °C, and orthoclase felspar, 0·000031 ($30·6 \times 10^{-6}$) per °C and both these minerals are commonly found together in many rocks. A very common example of insolation may be observed in the flaking of slaty stones present in an open coal-fire, a phenomenon technically called *exfoliation*. This insolation has a counterpart in the direct heating action of the sun on an exposed rock surface, the more so if the rock faces the sun's zenith position. The insolation effect is particularly marked in conjunction with rain and frost where water enters the cracks formed by exfoliation, freezes at night and hence splits the rock.

(*d*) *Weathering by wind*—Wind is particularly effective as an agent of erosion in hot, dry desert climates, causing sandstorms and the abrasion of hard rocks through the blast of wind-blown sand; this effect is especially noticeable close to the ground where the sand movement and eddies are densest. Wind causes also the removal of fine top-soil as dust particles and the formation of 'dustbowls' in regions of low rainfall. Dustbowls are more often the result of bad agricultural usage of land, and a notable example of soil erosion by wind was shown recently by the formation of the 'Bad lands of Nebraska', U.S.A. Such exhaustion of a virgin top-soil through over-cultivation after deafforestation and clearance of the natural vegetation, may take place rapidly. Intensive farming of cattle land for wheat on the plains of Kansas and Nebraska, U.S.A., had disastrous results within a period of 20 years and in the 'nineteen thirties' led to the formation of large desert areas. The same problems, although arising from a different cause, over-exploitation of sparsely vegetated grazing land for cattle breeding, affect East Africa today: also, in the 5 years up to 1963, a huge 'dustbowl' has been created in the virgin lands of western Siberia by Russian attempts to over-cultivate 90 million acres of unused earth for wheat production.

Remedial measures involving the accumulation of new top-soil are necessarily slow, men have learned by painful experience that careful use of land through afforestation, contour ploughing and correct rotation of crops, is essential to the environmental and ecological welfare of earth, animals and mankind.

The soil particles remain wind-borne until a barrier is reached over which the wind is unable to carry them, so that sand dunes (*Figure 2.3*) are formed as typical 'aeolian' deposits on sea shores and in deserts.

On coastlines, long mounds or ridges, approximately parallel to the shore and often 100–200 ft or more high, are formed by the prevailing inshore wind. These sand dunes tend to encroach inland as they migrate downwind and the usual method of fixation is to cover the dunes with tough, hardy 'marram grass' on their windward slopes, which prevents free movement of the grains. *Figure 2.3* indicates how the sand grains are blown by the wind up the windward slope and fall into the leeward trough at a steeper slope, fixed by the angle of natural repose of the grains at 30–35 degrees to the horizontal. The continuing action of the wind gradually moves a dune to leeward by this transference of sand particles from its windward to leeward slope.

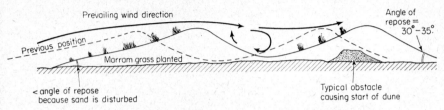

Figure 2.3. Formation and movement of sand dunes

Examples of top-soil formed from fine wind-blown material* of previous ages are to be found throughout the world today, the most famous being the *Loess* of China, a fine calcareous loam which is formed of material originally wind-blown from the Gobi Desert region. In Asia and Central Europe also there are thick accumulations of fine loam or dust blown from the Sahara region of N. Africa; mixed with humus and other organic matter such 'Loess' forms the famous Black Earth of the Ukraine.

Weathering by Living Matter

(*e*) *Organic action*—Plants retain moisture, thereby keeping the underlying rocks damp; their decayed matter in common with animal excretions, animal remains and other humus produce organic acids which attack these rocks, forming a rock waste susceptible to the eroding effects of rain, wind and runnels of water. The physical action of plant roots also encourages splitting of rocks. Although individual effects may be small, the cumulative result is appreciable.

2.4. DENUDATION—TRANSPORTATION—DEPOSITION

(*a*) *Ice and Glacial Action—Formation of a Glacier*

Snow forming the permanent snowfields, i.e. above the snow line in mountainous regions, seeps as water caused by the melting power of the sun; the water freezes again, however, under consolidation from the weight of surface snow or partly compacted 'névée' of snow with ice and, with the lower temperature prevailing at depth, forms an ice river beneath the less dense surface. The

---

* This material became anchored in pre-existing vegetation and buried it until the resulting mixture rotted to form a highly humus top-soil.

flow of ice is connected with the phenomenon of 'regelation', the effect of pressure in lowering its melting point. Regelation causes the compressed ice particles to melt at their points of contact, forming an icy water film which almost immediately freezes again (with the particles in a slightly different position) and giving them the appearance of a sliding or gliding motion relative to one another. Different movements can therefore occur in different parts of the ice mass, so that a glacier, while retaining its solid properties, moulds itself to the shape of existing land forms and the higher snows slowly push the ice, under the action of gravity down the mountain slopes into the valleys below.

'Crevasses' and cracks extending across a glacier are formed by its irregular motion, especially down steep inclines or over humps (*Figures 2.4* and *2.5*)

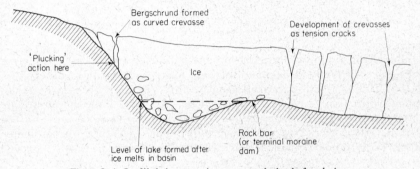

Figure 2..4 Ice filled cirque, corrie or cwm at the head of a glacier.

(A cirque is a C shaped hollow with steep walls found at the head of a glaciated valley and high up on a hill side. It has a level or gently inclined floor frequently occupied by a small lake (Llyn, Lochan, Tarn) when situated in the gouged-out rock basin)

Figure 2.5. Alpine mountain building. The Bernina Bergmassif (Note the rugged, folded character of the range, the frost/ice serrated arêtes (ridges) and the glacier in the foreground with crevasses)

(By courtesy of the Bernina Sewing Machine Coy.)

34

and tend to close up in hollows; the start of the separation of the glacier from the snowfield is marked by a large crevasse called a *Bergschrund*, which in common with the ice moves forward at a speed of several feet per day; e.g., 2–5 ft/day is an average value for alpine glaciers like the Aletsch, Rhone and Stubai. The erosion of a 'Bergschrund' often leaves behind it a steep-walled semi-circular amphitheatre called a 'Cirque' (the typical Cwm in Wales and Corrie in Scotland), formed by the plucking action in the lower levels of ice on the rearward rock face (*Figure 2.6*) over which it slides.

Figure 2.6. Llyn Cau—an almost perfect 'cirque'-form on the Cader Idris escarpment—N. Wales

(Crown copyright Geological Survey photograph. Reproduced by permission of the Controller, H.M. Stationery Office)

Glacial erosion and modification of land forms—Although ice is essentially rigid, brittle and crystalline, glacial flow is in effect that of a highly viscous fluid. A glacier, following the established land forms, swells out in broad valleys and contracts in narrow ones, breaking rocks from the floor and sides of a valley and carrying them along as 'moraines'. Side deposits are termed 'lateral moraines'; that in front of a glacier is called the 'terminal moraine', and at the intersection of two glaciers, a 'medial moraine' is formed (*Figure 2.7*); from the terminal moraine 'outwash deposits' are often carried away to lower ground by the terminal glacial streams.

Morainic material is carried mainly in the lower regions of the ice; there, the grinding of boulders and stones on each other or the valley floor produces a fine mud or rock flour, unweathered in the ordinary sense of chemical

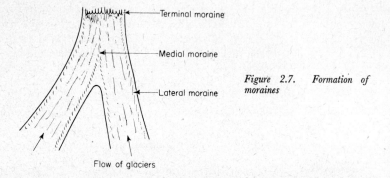

Figure 2.7. Formation of moraines

decomposition by rain-water. This material is also a relatively unsorted mixture of boulders and gravel or sand with clay, generally angular in shape, since its particles have not been rounded or graded like those of river-borne sediments.

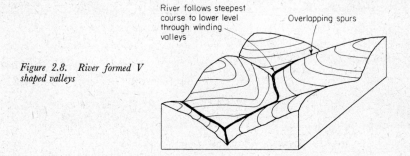

Figure 2.8. River formed V shaped valleys

Glaciers cut valleys from their river formed V shape to a U shape; the rocks over which a glacier passes are smoothed by the ice and gouged by ice-embedded rocks to give a 'glaciated', striated surface on their iceward slope, and a 'plucked' appearance on their leeward side. The same process occurs on the side slopes of a valley, leading in both cases to the appearance of *roches moutonnées* as smooth rocky hillocks composed of more resistant rock outcrops than their surrounding strata, projecting above the general level of the ground and its present-day vegetation.

Rock-bound basins gouged out of a solid rock bed are also a familiar feature of glaciated country. In Great Britain, Lock Coruisk in Skye, many tarns of the Cumberland Lake District and Llyns Glaslyn and Llydaw near

Snowdon, north Wales, are well-known examples of lake filled basins where the valley at the downstream end may be crossed on its 'rock bar' (*Figure 2.4*), which forms a natural dam for the impounded water.

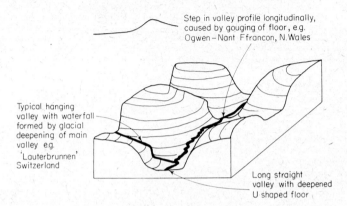

Step in valley profile longitudinally, caused by gouging of floor, e.g. Ogwen—Nant Ffrancon, N. Wales

Typical hanging valley with waterfall formed by glacial deepening of main valley e.g. 'Lauterbrunnen' Switzerland

Long straight valley with deepened U shaped floor

Figure 2.9. V shaped valleys modified to U shape by glaciation

Diversion of drainage—A glacial lake may be formed if the outlet to a valley is blocked by ice and water is thus impounded; its level may fall lower or disappear altogether if the ice dam melts, leaving behind 'gravel beaches' as evidence of former shore lines. The Parallel Roads of Glen Roy, Scotland, are probably the most famous example of this, with three distinct terraced beaches corresponding to successive levels of a former glacial lake, showing in clear view on the mountain sides of the Glen (*Figure 2.10*).

Should such an ice dam remain in position long enough for the impounded water to escape by new routes, or for existing rivers to become diverted through overflow channels, then some remarkable phenomena are left as evidence of the former glaciation. One important example in England, dating back to the 'Great Ice Age' (or Pleistocene Period), is the diversion of the river Severn through the Ironbridge Gap, from an original north-easterly course towards the Dee estuary to its present southward flow across Shropshire.

Characteristic glacial deposits—A 'terminal moraine' marks that place in a valley where a glacier's 'snout' remained stationary for a considerable period; here, the forward movement of the glacial ice just kept pace with the melting of its terminal face so piling up morainic material, until a somewhat faster retreat of the ice commenced when climatic conditions grew warmer. Material brought forward by the glacial flow thus collected in a large heaped mass extending across the valley like an earth dam; a much more rapid retreat of the ice, however, would have resulted in a more or less even spread of such deposits over the original valley floor.

37

Erratics—Single rocks known as 'erratics or boulder stone' may be transported hundreds of miles from their original source by ice and left anywhere. 'Perched blocks' of stone are often to be seen on steep hillsides or heights, where they were left attached to shaded patches of ice which remained

Figure 2.10. The parallel roads of Glen Roy, Inverness-shire

(The famous parallel roads are narrow terraces 40–50 ft in width which run horizontally along the mountain sides of Glen Roy and tributary valleys. The terraces occur at mean successive levels of 1148, 1067 and 855 O.D. and they mark the beaches and water-lines of a series of glacier-dammed lakes, their heights corresponding with those of three cols which in turn controlled the water levels.)

(Crown copyright Geological Survey Photograph. Reproduced by permission of the Controller, H.M. Stationery Office)

unmelted until a block had time to settle in an unusual or precarious position. When the base of a 'perched block' has been weathered away, mainly by wind, rain and frost action such an erratic may subsequently become a 'rocking stone', unusual enough to command much local interest as many do on the moorland heights of northern England (*Figure 2.11 Erratic*).

Boulder clay—Ground-up morainic debris, when consolidated, forms a stiff clay called 'boulder clay', derived from its parent mixture of crumbled soil and rock fragments and this clay includes pebbles as boulders of varying sizes; when unconsolidated, the same material is often called 'glacial till'. To the engineer, these deposits can be a most treacherous foundation material, because of their variable properties (both vertically throughout the thickness of a stratum and also over a large site area or whole region of their outcrop), especially the presence within them of large concealed boulders, easily mistaken for solid rock during borehole exploration for a particular sector of a major civil engineering work.

Boulder clay deposits fill the base of most valleys in upland regions of Great Britain and existing rivers are now cutting down through these drifts, often following the line of a former water course but also draining in entirely new directions* to those prevailing before glaciation occurred. On lowland areas, much drift extends beyond the original shore line outwards below

Figure 2.11. Erratic Block (ice transported boulder of mica-schist with tree growing in crack). Trefarthen, Anglesey.

(Crown copyright Geological Survey photograph. Reproduced by permission of the Controller, H.M. Stationery Office)

sea-level and in eastern England particularly, this led to the extension of land beyond the former pre-Pleistocene (i.e. pre-glaciated) coastline into the North Sea, when the sea level was much lower under the ice sheets, than it is today.

* An old buried channel of the River Mersey was encountered during construction of the Mersey Road Tunnel and shows that the flow of this river was completely reversed by the Pleistocene glaciation.

The Great Ice Age and glacial drift—It is, however, the vast ice sheets, like those of Greenland and Antarctica today, which during the Pleistocene period overspread and melted on lowland areas to leave the large deposits of glacial drift known to cover much of Great Britain and northern Europe at the present time. These deposits, mainly of 'boulder clay' varying in thickness from a few feet to a hundred feet or more, are accompanied by fan-like sheets of 'outwash gravels' which mark the southern limit of their former ice sheets. The latter deposits were laid down in the melt streams flowing away from the terminal surface of the ice (*see Figure 2.12*) and extend in England to the Surrey–Hampshire borders, well beyond the southernmost limit of glaciation in the Thames Valley (*Figure 1.4*)

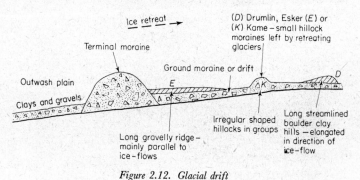

Figure 2.12. Glacial drift

(Behind a terminal moraine, it is usual to find groups of low hills or hummocks called Drumlins and low-winding ridges named Eskers as well as irregular knolls known as Kames)

Because the 'Pleistocene period' is a recent geological occurrence, the results of its glaciation are still very apparent in Great Britain, Europe and North America; the glaciated land forms of the northern hemisphere remain, as yet, largely unmodified by weathering action or erosion in many areas. The handbook* on British Regional Geology, N. Wales, gives a particularly relevant account of glacial erosion.

(b) River Action

The sediment a river carries in suspension is always eroding its bed and banks, especially so if it passes over a soft stratum and when, for example, the bed is steeper near its source where the river is young with a swift flow of 10 knots or more. This is a process known technically as *corrasion* which leads, ultimately, to the general overall lowering of a land mass. At the

---

* The *Short Guide* to the London Geological Museum of the Institute of Geological Sciences details all the available facilities and lists the various exhibits. There is a Reference Library of geological literature containing some 60,000 books, over 1,000 periodicals, 30,000 pamphlets, numerous maps and over 20,000 photographs.

present time, English rivers are lowering land at the average rate of 1 ft in some 4,000 years. The land surface of the Thames valley has apparently subsided about 15 ft since the years of Roman occupation, as evidenced by archaeological discoveries, but the total movement is not due solely to river corrasion. (*See* Isostacy, page 61.)

Modification of land forms—both valleys and gorges are cut by rivers, although a young active stream cuts a V section which only widens out into a U-shaped section as the stream gets older, when base and lateral erosion occur together. A river lives through stages of youth, maturity and old age according to its gradient and of the type of rock over which it flows. It cuts the characteristic V cross-sectional form in mountain regions, where the swift flow on a steep gradient engenders a powerful downward 'corrasion', through the rolling of large boulders and pebbles in part-suspension along the river bed; a broader and more mature form occurs in a valley tract where the gradient is easier and lateral erosion begins to equal basal down-cutting. Finally, when the river has graded the longitudinal profile of its bed into a smooth concave logarithmic curve extending from source to mouth, no further down-cutting takes place; sub-aerial erosion is active, however, on the valley sides and a saucer-shaped tract, ⌣, emerges which may be subjected to periodic floods involving more deposition of material than actual erosion. The open old-age river valley near base level becomes in the ultimate, an 'alluvial flat' or 'peneplain'* across which the river's sluggish water meanders as it approaches the sea and deposits material brought down in suspension.

The upper parts of a river are therefore young, or 'immature', and the lower reaches mature, while near base level (i.e., the sea) it may be termed 'old'; in general, the mature stretch of a river tends to extend gradually upstream as it ages. Most British rivers are mature but do not have the length or watershed area to compare with the great continental rivers of the world, like the Rhine, Danube, Volga; Amazon, Orinoco, Mississippi; Nile, Ganges and Brahmaputra.

River capture—where a river is very active at its source, it continually extends its valley inland and more water tends to collect in the resulting V; sometimes the ground of an adjacent basin is captured and its head-waters diverted from their original drainage directions to form tributaries of the more active stream.† The largest rivers are generally those which have captured most tributaries.

Waterfalls usually occur where a hard rock stratum or band crosses the bed of a river and, in interrupting the normal corrading process, forms a step in the river's longitudinal profile. A waterfall tends to move backwards

---

* Meaning 'almost a plain'.

† A classic example is shown by the headwaters of the River Blackwater which has been captured by the River Wey near Farnham, Surrey. (*See* pp. 77–78, The Wealden District.)

against the stream with time as the river erodes even the hard stratum, so forming a gorge. (*Figure 2.13.*) Thus the Niagara Falls have moved back

Stages in formation and retreat of falls
1. Vertical corrasion of river removes soft rock below face.
2. Swirl and spray of water undercuts hard rock stratum—overhanging ledge collapses and the falls retreat.
3. Undercutting so great that overhang collapses through weakened vertical joints.
4. Process recommences on new face.

Hard rock stratum; igneous, limestone, sandstone

Overhanging ledge collapses and falls retreat

High force, Teesdale; Niagara, typical

Softer rock (e.g. shale, clay) undercut by river: often possible to walk beneath fall

Figure 2.13. Waterfall in horizontal strata (e.g. Niagara. *See also* Figure 2.15, Thornton Force, England)

Figure 2.14. Potholes. On River Taff, below Craig-yr-Esq near Pontypridd
(Crown copyright Geological Survey Photograph. Reproduced by permission of the Controller, H.M. Stationery Office)

265 ft in 63 years and are now receding at a rate of some 12–15 in annually.* At Niagara, a hard limestone stratum overlies softer rocks; the waters from Lake Erie flow over the falls into a gorge, now some 7 miles long, which

* The rate has slowed since the construction of hydro-electric power plants and the consequent diversion of some water flow from the main stream.

has been formed by the retreating falls since the end of the Great Ice Age 10,000 years ago.

Another phenomenon occurring due to the action of rivers flowing in hard strata is the formation of 'potholes' (*Figure 2.14*). This is caused by frictional and centrifugal erosion of rock layers, by the circumferentially moving sand and pebbles trapped in crevices or hollows of the river bed in swirling water; e.g. as at 'Devil's Bridge' near Aberystwyth in Wales.

Both valleys and gorges are cut by active rivers and, although in hard rock a gorge may be formed, in weaker rock a river valley widens out as more lateral than base erosion takes place, assisted by the normal sub-aerial weathering on the valley sides above the water level of the river.

The Grand Canyon, Colorado, in places over a mile deep, and in all, 330 miles long, is a remarkable gorge phenomenon caused by the rapid vertical corrasion of the regenerated and active Colorado river in uplifted and relatively soft horizontal 'Paleozoic' sediments. The generally arid climate of this region has prevented the usual lateral weathering on the steep sides of the gorge. This would normally have occurred through constant rain action and the gorge would have become widened to a complete valley form. *Figure 2.15* shows a typical valley form.

Figure 2.15. Thornton Force, Ingleton, Yorkshire
(Crown copyright Geological Survey Photograph. Reproduced by permission of the Controller, H.M. Stationery Office)

Meanders—'Old' rivers flowing slowly at perhaps 2 knots or less, 'meander', since they tend to corrade laterally rather than downward and do not have the power to remove direct obstructions. At a bend, coarse material is dropped out of suspension to form a beach on the convex side of the stream, where the flow is reduced, whilst the centrifugal effect of the water flow causes a higher velocity and hypernormal erosion on the concave side. These two effects together sharpen the bend, increasing yet more the circumferential erosion, and if this cumulative process is particularly exaggerated, an 'Oxbow' and ultimately a 'Mortlake' may be formed (*Figure 2.16*).

Stages in formation
1. Flow diverted by landslip or obstruction.
2. Concavity of outside bank increasing; deposition on inside convex bank; both are accentuating the curve.
3. Development of oxbow loop with neck continually narrowing.
4. River breaks through neck during flood flow leaving mortlake, which in time fills with sediment.

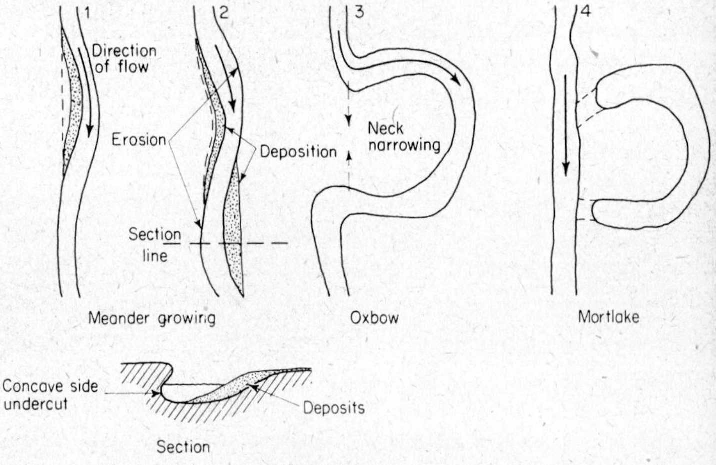

Figure 2.16. Development of meanders

It should be noted that the carrying power of a stream varies at the rate of the fifth or sixth power of its velocity. If the velocity is trebled during a flood period, the carrying power is increased several hundred times, and even large boulders may be rolled along the bed, or substantial particles of rock carried in suspension. Some 800,000,000 tons of solids are carried annually into the sea by the rivers of the U.S.A.; the Mississippi alone removes 400,000,000 tons of solid matter per annum, of which one-quarter is in solution and almost all the rest is in suspension. When its water velocity is checked, however, the carrying power of a river is sharply reduced and much transported sediment is almost immediately dropped. On emerging from a constricted V type mountain valley into an open valley or plain, where its velocity is checked by the reduced grade and wider form, a river builds up an 'alluvial cone' by deposition of mainly coarse particles. This self-produced cone may eventually so interfere with the flow as to cause the river to change course and start a 'meander' which, by its natural growth, must further widen the lower valley or plain.

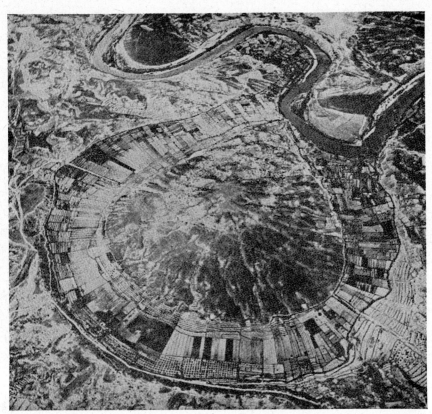

Figure 2.17. Oxbow of fertile land in a rocky barren region of Spain. At the neck, a new channel has been specially cut for the river Ebro and the mortlake so formed is now a fertile valley among barren hills. The central area surrounded by the dry meandered bed shows radial dendritic drainage

(By courtesy of Paul Popper)

Alluvium is a fine grained silty and clay material with streaks of sand deposited in geologically recent times by rivers at base level near their mouths, or over their 'flood plains', as the alluvial plains or 'flats'* of open old-age river valleys, in which meanders form, are often completely flooded. The river normally follows a narrow channel between 'levees' or banks of coarse sandy material deposited just outside the flow stream of its main current. During flood periods the river overflows its 'levees' and from the

* *See* Section 2.8, page 64, *Rivers and Waterfalls.*

stagnant flood water, particles of silt, mud and clay are deposited, thus raising the general level of 'alluvium' already spread on the plains; from the slight flow remaining in the main channel, coarser material is also dropped, which further raises the 'levees' (*Figure 2.18*). 'Alluvium' is often a very thick silty deposit of variable quality with which the foundation engineer has to contend. It causes difficult construction problems in many important cities or towns situated near river mouths and estuaries or on a widespread flood plain.*

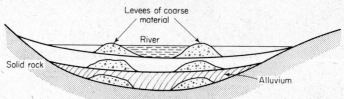

Figure 2.18. Development of Levees on a flood plain

Uplift of land—rejuvenation—incised meanders—Uplift of land mass can lead to 'incised meanders', by causing the 'rejuvenation' of a mature river, which is then forced to begin a regrading of its bed to cut down again towards the lower relative base level, while maintaining its established course. The deep winding valley of the River Wye below Ross and the gorge of the Wear at Durham, are both classic examples of this phenomenon. 'Incised meanders' are always evidence of uplift when the features of a mature river have been, or still are, combined with those of a young and still active stream.

Gravel terraces—An old river cutting downwards again after rejuvenation, and following an elevation of its surrounding land, may also leave 'gravel terraces' on the original but now higher valley sides and at a regular height above its most recently established flood plain.

The river cuts a new and narrower valley through existing alluvial deposits while following its mature course on the already wide flood plain and, later, in a renewed old age, deposits more alluvium at the lowered base level to form another of these plains. In the Thames valley, evidence in the form of flints, bones and implements, exists on several 'terraces' of well-drained gravel, e.g., Boyn Hill and Taplow*, which mark the sites of early settlements chosen by men, and these terraces give proof of the successive uplifts of land in south-east England during former ages.

Deltas—When a river flow is checked at its entrance to the sea, or to a lake, a shallow water 'delta' of stratified gravel, sand and clays is often formed as the water velocity progressively decreases. The coarser particles are dropped directly from suspension in order of their grading; where a river enters the sea the common estuarial muds, consisting of silts and clays (both materials

* *See* page 199 and I.G.S., Handbook, 'London and Thames Valley', pages 49–52.

formed from very fine particles), are caused by the phenomenon known as 'flocculation'.* This, following from the neutralization of static electricity on their ionized particles in contact with salt water, means that thereafter such mud banks or flats remain remarkably stable, despite the wash of a strong water flow during flood or ebb tides; by contrast, lake deltas always consist of clean sandy sediments. A 'current bedding effect' is often apparent in shallow water deposits and is caused by deposition of sediment in layers from a current flowing in one direction at varying speeds—followed later by deposition from a similar current flowing in another direction—usually after the washing away of the upper part of the earlier formed layers has occurred, to leave a truncated appearance in the lower deposits at their upper boundary surface. The effect often extends through large thicknesses of deposits, even after their compaction to rock, and is always of the form shown in *Figure 2.19*; the sandstone layers in the rock walls of the Grand Canyon, Colorado, are a particularly rich source of visual evidence for this 'current-bedding' phenomenon.

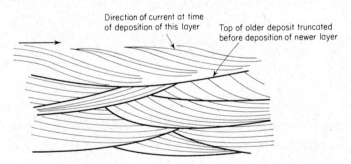

Direction of current at time of deposition of this layer

Top of older deposit truncated before deposition of newer layer

Figure 2.19. Current bedding effects in fine silt—vertical section

The formation and growth of 'deltas' is comparatively rapid. Famous lake deltas, such as those at the head of Windermere, Derwentwater in Cumberland and Lac Léman in Switzerland, where the river Rhone enters the lake, have all produced flat fertile tracts of valley in mountain regions. Deltas of major rivers like the Rhine (which forms the bulk of South Holland), the Ganges and Mississippi, cover huge areas, although powerful tides may, in certain instances, prevent the further growth of deltaic deposits.

Submergence of land may convert river mouths and deltas into open estuaries, while tidal effects working upstream can greatly change the nature of a river's lower reaches unless the river is 'trained' for navigational purposes. Most tidal rivers used by shipping, like the Thames and Mersey,

* By 'flocculation' is meant that these particles which would ordinarily remain in suspension until far out at sea (*see Figure 2.20* and *Marine Deposition*, page 54) then coagulate into large enough lumps for their deposition to occur at the same water velocities that lead to sedimentation of the sand/silt fractions.

are now constrained to flow between embankments, training walls and in dredged channels.

(c) Sea Action

(i) *Erosion by waves*—The direct mechanical effect of wave momentum and wave carried débris is greatest on a steep shore line where the waves break close inshore; recurring or periodic impact pressures of up to 3 tonf/ft^2 may occur in gales and severe storms.* On a flat sandy shore, little damage is caused as the waves break a long way out, or their energy is absorbed by the friction of beach material and rapidly diminished. Attack by waves is therefore very varied on a coastline, the weaker points being rapidly eroded while little effect is seen elsewhere. Headlands and cliffs are thus formed as hard rock *outcrops* jutting out of a softer shore line which elsewhere has been more extensively eroded. The English coast is being eroded, mainly in the East and South, by the sea at an average rate of one foot annually. This erosion involves an amount of land material about equal to that removed by the agents of sub-aerial weathering alone acting on exposed surface rocks inland.

Where the outcrops of rock *strata* run parallel to the shore, a coastline of little variation and composed of one rock type, results; where the outcrops cut across the boundary of a coast, however, a wide variety of rock types will be exposed, and their relative hardnesses influence the shape of the resulting coastline which then becomes varied and irregular.

Stacks, platforms and caves†—In any near horizontal bed of softer rock at the base of a cliff, long low caves may in places be hollowed out of the rock by sea waves; two or more such caves cut through a headland from either side, may meet to form a tunnel with a natural arch roof; the sea also enlarges joints and bedding planes in the cliff and tunnel walls, and ultimately, the tunnel roof collapses, leaving a rock 'stack' detached from the main cliff face.

* *See* References marked with an asterisk, pages 67–68.

† The references given below relate especially to the study of features mentioned in the following sub-sections.

Geological and O.S. maps for Chesil Beach, Lulworth Cove and Stair Hole, Dorset, etc.: 1 inch geological map sheets—Ref. 341–2–3 and 1 inch corresponding sheet memoirs; Diorama photo MNL 513 and topographical 1 inch O.S. map.

Orfordness: Ref. 1 inch geological map sheet 162; regional handbook for geology of East Anglia and 1 inch O.S. maps; Photo A6940.

Needles, Isle of Wight: Photo MNL 501. 1 inch drift geological Isle of Wight Special map.

Cornish coast and platform: Photos A235, A570; Ref. 1 inch geological map sheets, 351, 358–9.

Seven Sisters, Sussex: Photo A5414; 1 inch mémoir for map sheet 359.

Regional Geological handbook, S.W. England and 1 inch O.S. maps.

Regional Geological handbook, The Wealden District and 1 inch O.S. maps.

Regional Geological handbook, The Hampshire Basin and 1 inch O.S. map.

(For full details of the availability of all such references—*see* Section 2.8.)

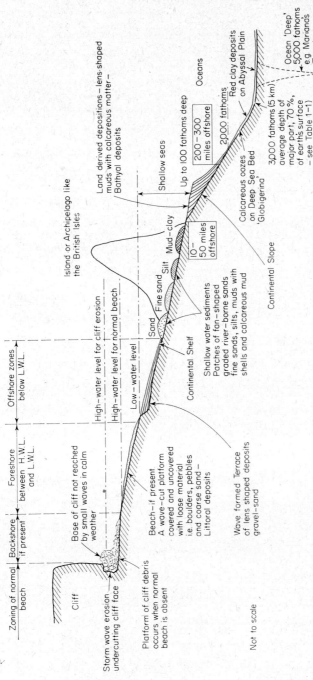

Figure 2.20. *Types of beach and sea-bed formation*

Zoning of normal beach

Backshore if present | Foreshore between H.W.L. and L.W.L. | Offshore zones below L.W.L.

High-water level for cliff erosion
High-water level for normal beach
Low-water level

Cliff

Base of cliff not reached by small waves in calm weather

Storm wave erosion undercutting cliff face

Platform of cliff debris occurs when normal beach is absent

Beach—if present
A wave-cut platform covered and uncovered with loose material i.e. boulders, pebbles and coarse sand — Littoral deposits

Wave formed Terrace of lens shaped deposits gravel—sand

Not to scale

Continental Shelf

Shallow water sediments
Patches of fan-shaped graded river-borne sands fine sands, silts, muds with shells and calcareous mud

Sand   Fine sand   Silt   Mud—clay

10—50 miles offshore

Continental Slope

Calcareous oozes on Deep Sea Bed 'Globigerina'

Red clay deposits on Abyssal Plain

Island or Archipelago like the British Isles

Land derived depositions — lens-shaped muds with calcareous matter — Bathyal deposits

Shallow seas

Up to 100 fathoms deep

200—300 miles offshore

2,000 fathoms

3,000 fathoms (5 km) average depth of major part, 70%, of earth's surface — see Table 1-1

Oceans

Ocean 'Deep' 5,000 fathoms e.g. Marianas Trench

Note: Mid Atlantic Ridge: World's greatest mountain range

Against a cliff face, a 'wave-cut platform', is often formed at low-water level, due to the under-cutting of its base by wave action; this undercutting followed by continual sub-aerial weathering above causes a collapse of the

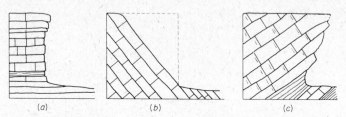

(a)   (b)   (c)

Figure 2.21. Forms of cliff in relation to rock structure (a) horizontal strata with vertical jointing (b) jointed beds dipping seawards (c) undercut cliff in rock strata dipping inland

over-hanging cliff face into a heap of debris. It is thus clear that the 'platform' will consist of the larger debris from the base of the cliff lying between the high- and low-water tide levels which the sea is unable to remove (*Figure 2.20*).

Figure 2.22. The Stacks of Duncansby, Caithness

(Cliffs of old red sandstone, with stacks of the same rock on a platform of marine erosion)
(Crown copyright Geological Survey photograph. Reproduced by permission of the Controller, H.M. Stationery Office)

In hard rocks with horizontal *bedding* and vertical *joints* a sheer cliff may be formed, showing that locally the marine erosion is more rapid than any

Note: For definition of italicized terms, *see* Chap. 4.

atmospheric weathering. If there are other joints, the sea enlarges them to form 'blowholes' or 'caves', with wave impact attack leading to the ultimate collapse of a cave roof and the rock above; in this, the compressed air effect of wave movement in a confined space plays the larger part in separating the rock layers surrounding the hole or crack.

When rock layers *dip* steeply seaward, their tendency to slip along their natural '*bedding planes*' determines the landward inclination of any cliff formed: conversely, when the rock layers dip landward, cliffs may form with considerable overhang above sea level (*Figure 2.21*).

*Figure 2.23. Erosion at Seaford Head on the Sussex Coast near the Seven Sisters**

(18–24 in is cut back yearly. Rain washes turf away from 283 ft high chalk cliff top. Winter frost fractures more rock from the cliff face. The crane is working on the sea defences for Seaford town at a spot once covered by the cliff, which has lost over 50 ft in the last 30 years).

(By courtesy of The Guardian)

The destructive power of shingle and sand in waves to 'comminute' (i.e., break up) rock at a cliff foot, or to damage sea defences, is also well known— the smaller debris so formed being removed by the undertow of receding waves. To protect the land against erosion on soft coastlines, sea walls, timber stakes or piles and groynes are used (*see* pp. 52–53, in relation to 'littoral drift'). Much information relating to the coasts of Great Britain was published by the 'Royal Commission on Coast Erosion' between 1909–11* and

* *See also* footnotes to page 48.

evidence of a high rate of marine erosion may be seen by comparing old and new editions of Ordnance Survey maps* for a particular locality like Seaford.

Landslides or slips are the movements of rock masses caused by the combined action of weather, ground water, and gravity on unstable materials and are similar in occurrence to the fall of cliff faces. They frequently occur near sea coasts or adjacent to the shore line and contribute to the breakdown of cliffs by wave action.†

Littoral drift of shore materials, i.e., the removal of debris by wave action and gravitational undertow along the shore by coastal or tidal current, takes place on most beaches (*Figure 2.24*). Timber stakes, piles, and concrete or

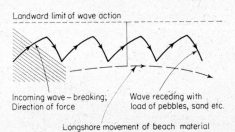

Figure 2.24. Littoral drift

wooden groynes running seawards, are constructed principally to obstruct this type of movement; they slow down erosion on soft coastlines, since, as mentioned previously, a good beach affords the land considerable protection from wave action. Indeed, without the removal of debris, beaches are built up and stabilized so that any damaging wave erosion is largely eradicated.

Material may be carried by a sea current for long distances until it comes to rest in the lee of a headland, at a bend in the coastline, or meets the contrary flow of water issuing from a river mouth where the current is checked; then, on that side of the river nearest to the origin of the material, the material will be deposited in the form of a 'shingle spit' which in time, may grow to a substantial size, dependent upon the relative strengths of the longshore sea currents and outward river currents, or the presence of further obstructions. On the east coast of England, Spurn Head at the mouth of the Humber estuary‡ and on the south coast, the Chesil Beach,§ a long storm-formed promontory near Portland, are both outstanding examples of the type of boulder and gravel 'spits' so formed. Often the mouth of a river is completely diverted, especially when the land-borne sediment it carries is piled up behind a growing spit. In Suffolk, the River Alde flows eastwards and then for some 10 miles southward from the town of Aldeburgh situated

* *See* Question 4, Section 2.7 and map sheet 334 and also footnote, page 48.
† *See* London Geological Museum Postcard A9763 of Folkestone Warren Landslip, Kent and Hutchinson (1969).
‡ Geological map (Reference No. 81).
§ Map (Reference No. 342).

at its original mouth; the river channel now flows parallel to the coast but deflected from its former mouth behind a 'long-shore spit' known as the 'Great Orfordness Spit'* (*see Figure 2.25*). When two opposing sea currents meet, a 'bay-bar' and 'foreland' of shingle may grow out seawards across a bay, thus affording protection to the gradual formation of mud flats and salt marsh behind it. The formation of such a bar has transformed the Dovey estuary in Wales into just such a region of sand banks and salt marsh, and the town of Aberdovey, once a thriving port, has lost much of its earlier importance as the River Dovey's channel is now small, narrow and tortuous.

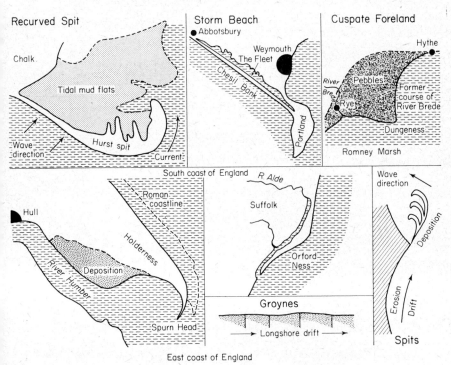

Figure 2.25. *Examples of marine and river deposition on coastlines*
(From P. T. Silley, *Physical Geography*, by courtesy of Schofield and Sims Ltd.)

Various similar examples indicate that the prevailing drift on the south coast of England is eastwards and on the east coast, the direction of littoral drift is predominately southwards. The drifts in conjunction with the coastal currents formed by the rise of tidal water in the shallow North Sea and English Channel, result in deposition of shore material at the famed Goodwin Sands, the 'graveyard of mariners' near the Straits of Dover.

* Geological map (Ref No. 162). *See also* footnote page 48 on Stacks, Platforms and Caves.

On the west coast the prevailing long-shore drift is northwards in the shallow Irish sea, and these various directions of drift are probably determined by the movements of shore material due to the severest storms occurring around British coasts throughout the year. In addition to the information available from the publications already listed, and indicated by asterisks on page 67, fuller descriptions of the means of coast protection, outlining remedial engineering works constructed on the shores of Great Britain and Holland, are readily obtainable (e.g., from the Library of the Institution of Civil Engineers).

(ii) *Marine Deposition* (*Figure 2.20*)—All denudation from the land eventually finds its way into the sea, forming either 'littoral' or deep sea deposits. Littoral deposits, which as we have noted lead often, through long-shore drift currents, to the addition of new land on some parts of a coast, contain mainly graded gravels and sands and form in the shore zones as sediments collected between high- and low-tide levels. As already instanced, when river-borne material accumulates behind 'spits', the formation of areas of low-lying marshy land is encouraged; the forelands of Romney Marsh and Dungeness* on the south coast (*Figure 2.25*) and the Norfolk Broads in East Anglia have all partly evolved in this manner.

Shallow water deposits settle at depths from below low-tide level to about 100 fathoms on the gently-sloping 'continental shelves' of major land masses and include fine sands, silts and muds with embedded fossils, shells and other organic matter.

Deep sea or bathyal deposits, which are laid down on the sea-bed regions of 'continental slopes', beyond the confines of the 'shelves' and the 100 fathom water line, consist of fine clays and oozes deposited out of slow-moving currents: deep sea muds, similarly deposited out of near stationary water between the depths of 500 and 1,500 fathoms, are coloured blue, green or red in order of depth and are the finest land derived materials.

'Deep sea oozes', or 'abyssal deposits' include Globigerina, a calcareous deposit, formed from sea organisms and found on the Atlantic, Pacific and Indian Ocean floors at depths down to 2,000 fathoms; this type of deposit is estimated to cover 50 million square miles of sea floor. 'Radiolarian' is a similar siliceous deposit found in the South Pacific Ocean at greater depths, but of smaller extent, covering about 3 million square miles in area. 'Red clays' also occur from 2,200 fathoms downwards to the greatest depths and also cover approximately 50 million square miles of ocean bed. The sedimentation of bathyal deposits often shows 'current bedding effects', more particularly so for those land derived, according to the general current directions at their time of deposition (*see Figure 2.19*) as described previously under 'Deltas'.

(iii) *Effects of emergence and submergence of land*—As already mentioned in regard to river estuaries, changes in sea level relative to the land, whether

* Geological maps (Ref. No. 319–320).

caused by uplift of land (or a comparable fall in sea level) and sinking of land (or conversely a rise in sea level), have a pronounced effect on a coastline. The reasons for such a change are not always apparent, but some ways in which it might occur are explained by the 'theory of Isostacy' (*see* page 61) and illustrated by the effects of the pleistocene glaciation on former land and sea levels. Results are most spectacular when an irregular mountainous coast has become a *coast of submergence*; the drowned glaciated valleys produce long steep-sided and narrow indentations known as *fjords* in Norway, or *lochs* on the west coast of Scotland. Such inlets are often up to 100 miles long and their winding pattern together with continuations into branch valleys indicate a continuity of land and submarine topography: the glacier-scored depths of the Sogne fjord which cuts for 112 miles inland through the south Norwegian mountains to near the Jostedalsbreen, Europe's largest glacier, has rock walls towering 5,000 ft above M.S.L. Such deep fjords, through which the largest liners can sail, have been much used as waterways for ocean going vessels since Viking times.

Drowned river valleys, while broader or shallower in appearance and, therefore, not visually so striking as fjords, may be seen on the south coast of Cornwall and Devon, at Falmouth and Plymouth, or at Milford Haven in Pembrokeshire, south Wales.

Submerged forests are usually found on gently shelving shores, where they appear as wave-worn tree stumps at low water mean tide and are clear evidence of relative subsidence. On the west and south coasts of Britain examples are to be found in the Mersey and Bristol Channel, and elsewhere in Devon, Cornwall, Wales and Scotland.

A coast of emergence on the other hand, produces a simple straight gently-sloping coastline in a relatively soft unvarying rock, marked by features of marine action at points above sea level, such as *raised beaches (Figure 2.26).*

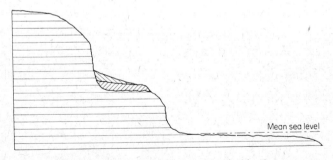

Figure 2.26. Sketch section through a raised beach

In Great Britain these deposits are found in many coastal areas and consist of shingle and/or sand mixed sometimes with sea shells, positioned at heights of 25–50 ft above existing sea level. *Raised beaches* similarly are the sign of an uplifted wave-cut platform. *Emergent coasts* are commonly featureless, harbourless and surf ridden until marine erosion slowly develops bays; the west coast of Africa and the Gulf of Florida, south-eastern U.S.A. are typical present-day examples of such shorelines.

The Denudation-Deposition Cycle

The consequences of the processes of denudation and deposition may be likened to a closed chain reaction which is continually in progress and which can be illustrated diagrammatically in the following manner:

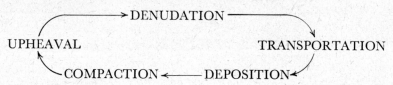

The resultant effects of the denudation-transportation processes may now be considered in detail and summarized. In deposition, the layered sediments formed by river transportation are, we have learnt, sorted in order of their particle sizes into gravel, sand, silt and clay (*Figure 2.20*); apart from deposition in the river itself, the coarser particles are dropped inshore as the carrying power of any river falls sharply when its flow velocity is checked on entering a sea or lake. The finer particles are normally held in suspension in a current until far out at sea, but fine particles are often trapped in between coarser material and the formation of *estuarial mud* by the *flocculation* of colloidal particles, mentioned previously, is also an exception to the general rule. It should be remembered also that water-borne material is rounded by knocking and bumping, the larger particles being rolled along the bed of a stream and the finer particles being held in suspension by eddies.

The various types of 'mechanical sedimentary rock' so formed are best classified by grain size as in Table 2.2.

Table 2.2. Mechanical Sedimentary Rocks

Grain size	Unconsolidated	Consolidated	Remarks
>200 mm 200–60 mm 60–2 mm 2–0·6 mm 0·6–0·2 mm 0·2–0·06 mm 0·06–0·002 mm <0·002 mm	Boulders Cobbles Gravels Coarse ⎤ Medium ⎥ Sand Fine ⎦ Silts Clays	⎡Conglommerate ⎣Coarse Medium ⎤ Sand- Fine ⎦ stone Siltstone ⎡Mudstone ⎢ (no lamination) ⎣Shale (laminated)	Water-borne particles are more or less sorted during transportation and deposition. Classification is that commonly used by civil engineers.

The deepest oceanic deposits like the 'Red Clay' mentioned previously are derived from very fine volcanic and meteoric dust which settles infinitely slowly from the earth's upper atmosphere, whereas the majority of fine clays and organic oozes also accumulate extremely slowly under water, but do so not entirely beyond the confines of 'continental shelves' and 'slopes'.

Chemical Sediments (*see* Table 2.1)

There is also the phenomenon of the precipitation of 'chemical sediments' from solution, when evaporation of water through the sun's heat in sub-tropical and equatorial latitudes is greater than that allowable for the concentration of salts present in their natural solution. Common salt (sodium chloride, $NaCl$) is thus deposited, chiefly in the shallow water of land-locked seas and salt lakes, or by 'desiccation' from ground water in enclosed salt pans, wherever such excessive evaporation is possible in the earth's arid regions. Other common chemical deposits of this evaporite type are rock salt (also $NaCl$ but in its crystalline form), 'anhydrite' and 'gypsum' (forms of calcium sulphate, $CaSO_4$) and the chlorides and sulphates of potassium and magnesium. The term 'chemical', however, includes such deposits as 'replacement' rock salts formed by chemical reactions out of solution in water; organic phosphates like the bird-originated 'guano' of Galapagos and Chile and nitrates such as 'saltpetre', as well as the carbonate limestones like dolomite and travertine of inorganic origin and various lime ooliths ($CaCO_3$). Many limestones, of course, like their cementing oozes, are derived from the layers of decaying shell organisms on a sea bottom and are thus of organic (e.g., shell or coral) origin. Over 2,000 million tons of salts, the result of 'chemical weathering', are transported by the world's rivers annually into the sea, but their concentration, consisting principally of sodium chloride ($NaCl$) remains very small (35 parts per 1,000) and re-markably constant, since any dissolved calcium carbonate is continually used up to form shelly limestones and corals by the lime-secreting sea organisms and plants.

Sedimentation effects; Bedding planes; Joints; in situ *Deposits*

Thus at various depths and for countless aeons of time, ever-thickening layers of sedimentary materials accumulate with their accompanying fossil remains. The sediments in such a region of deposition change slightly or sharply when changes in sea currents, movements of the sea bed or of the material carried to that locality by rivers, suddenly occur; hence a bed of sand may be covered by a bed of chalk and other regular alternations of materials may occur.

When the same material accumulates without a break for a long period there will be evidence of horizontal layering (e.g., the grain of rock). *Bedding planes* representing the former surface of a sea-floor, may either be far apart, producing the 'freestone' (*see* Chap. 3) of quarryman's jargon, or they may be so close as in 'shale', that there are as many as fifty laminations to the inch. All beds represent the process of stratification and lamination; the 'freestone' is an outstanding example of the former; shale is a notable example of the latter. Each lamination or bedding plane represents a break in the deposition of the material, although the time involved in the break, may be far greater than the period taken in deposition.

The increasing pressure of overlying material on the earlier formed sediments causes compaction of these layers, thus squeezing out some of the water (connate water, *see* Chap. 6) held captive in the interstices between their particles, until, for example, a 'mud' (containing say 10–20 per cent solid and 80–90 per cent water) becomes a 'plastic clay' with a 30 per cent moisture content. Further compaction squeezes out most of the remaining connate water and the rock becomes a 'mudstone' or a 'shale'. The rise in temperature due to the blanketing effect of the superimposed layers, and the *temperature gradient* established through the earth's crust, causes chemical changes and re-crystallization of minerals, so leading to the formation of the low-temperature sedimentary rock types. In the case of porous rocks, 'cementation' by silica, calcium carbonate and silicon-iron oxides occurs, as these substances are re-deposited from solution in the interstices of the rocks during the squeezing-out process.

This compaction, shrinkage and drying of its sediments contributes to the development of cracks and joints through a rock, more or less in the vertical direction initially (*see Figures 4.1* and *4.10*). However, movements in the earth's crust, of which the sedimentary rock is now truly an integral part, cause other lateral joints to form, which at depth remain closed, but on exposure to weathering processes at or near ground surface become open fissures. Good sets of natural joints are of great assistance in quarrying but frequent irregular joints, due to the intense stressing of earth movements, make a stone worthless for building purposes except perhaps as hardcore, ballast or concrete aggregate.

To complete the picture, mention should be made of some less frequent types of surface rock or sediment, derivatives which are the untransported products of weathering; i.e., fragmental and chemically weathered rock deposits left *in situ*, above their solid parent rock. For example, the clay-with-flint layer covering the Chalk Downs in England; 'Laterites' of the tropics; 'bauxite', the aluminium ore; even igneous lava fragments or 'pyroclastic' rock, comprising particles and dust blown out of a volcano or vent to settle on the ground, are considered to form a localized sedimentary deposit after their passage through the air.

2.5. METAMORPHISM

Certain circumstances may cause sedimentary and igneous rocks to be deeply buried again, re-compressed and subjected to high temperatures or pressures, a process known as *metamorphism*. Their minerals then in turn, like igneous rock minerals at the earth's surface, become unstable and new minerals form or reform from the solid state under these conditions of high temperature and/or pressure of a kind prevailing several thousand feet below ground surface. Pure limestone re-crystallizes to marble; sandstone to quartzite; coal to graphite; granite to gneiss, and clays and shales are transformed into slates. Older metamorphic rocks, which frequently contain localized rich concentrations of both common metallic ores and precious metals, are widespread throughout the earth's crust as parts of the eroded remnants of ancient mountain chains with their roots criss-crossing the present-day continental regions.

Metamorphism is thus produced in two ways. In *thermal (or 'Contact') metamorphism*, molten igneous rock magma is intruded through cracks and fissures of the crust; the surrounding rock is re-crystallized by the intense local heat, leading to the formation of a 'metamorphic aureole' (*see* Chap. 4) around the intrusion. *Pressure or dynamic metamorphism* occurs on a regional scale, usually in those areas affected by mountain building, and is created by enormous earth pressures as well as high temperatures during major re-adjustments of the earth's crust. Such metamorphosed rocks, whether igneous or sedimentary in origin, usually show 'foliation planes' (i.e., lines of crystal flow), perpendicular to the main stress direction, giving a distinct 'grain' to the rock. This foliated structure is generally most apparent in the schists (a fine-grained layered rock), and gneisses (a coarse wavy-grained speckled rock), or when slate has been transformed into a fine-grained schistose rock.

Thermal metamorphism commonly produces speckled rock types like the hornfels, and high density minerals are common in all metamorphic rocks. This is due to their reduction in volume which occurred under the conditions of high pressure or restricted expansion prevailing during their formation, and also by the addition of silica and iron to their mass. Fuller descriptions of the rock types mentioned above and areas of their occurrence follows in Chaps. 3 and 4.

There is thus a further cycle of events working within the broader denudation-deposition framework outlined previously, a cycle which can be illustrated thus:

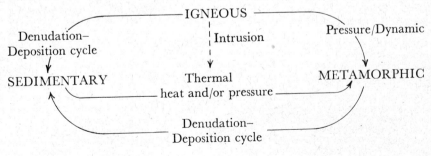

Exposure of metamorphic rocks to surface weathering brings instability to their mineral content, as with the igneous and many sedimentary types, and leads again through the processes already described as causing the formation of new sedimentary rocks.

2.6. THE GEOSYNCLINE AND ITS RAMIFICATIONS

Earth Movements and Isostacy

What are the circumstances then which cause the earth movements and pressure build-up of such a magnitude to bring about a rock metamorphism of the type already described?; also, what forges the last link, 'upheaval', in the chain of cause and effect required to bring the denudation-deposition cycle back to its starting point and a recommencement of the denudation processes on newly exposed rock?

Throughout the earth's history, time after time, land has sunk below the sea or a former sea floor has emerged to form new land. However, really major crustal re-adjustments of this type (called mountain building, 'oro-genies' by geologists) take place only after the prolonged and continuous accumulation under seas and oceans of up to 40,000 ft of sediments—a total thickness of approximately 8 miles—which equates to that of the earth's under-ocean basaltic crust and parts of the Sial; during this process, the sinking of the sub-oceanic Sima crust into the upper mantle must occur to permit such an accumulation of sediment in a vast but relatively shallow trough-like depression called a *geosyncline*. This is usually a long narrow area; the Mediterranean region is a typical geosynclinal basin and the European Alps originate from earlier sediments so formed there. The maximum down-warping of the earth's crust which would appear to be necessary to accommo-date the accumulated sediments must inevitably lead to revolutionary changes, as the sides of the depression move inwards with a vice-like action, sometimes after a settled period of perhaps 100 million years. Extreme faulting and folding of the sediments and outer crust builds mountain chains like the Himalayas, or the Alps, as enormous stresses are relieved deep in the earth's interior crust; thrusts also occur in which masses of rock are driven almost horizontally for many miles while overriding their underlying strata and leaving evidence of huge lateral, as well as vertical displacements, on the thrust planes (e.g., Moine Thrust in Scotland).

Many geophysicists calculate that seven such major orogenic readjustments have occurred during the last 1,500–2,000 million years of earth history, associated with seven recognized geological epochs of mountain building since the time that the earliest known sedimentary rocks were formed. Four orogenic epochs have definitely been allowed for in the record of British geological history, the earliest being extremely difficult to decipher, and this may account for the dim obscurity of still earlier occurrences.

The cause of these vast phenomena is thought to be due to the viscous liquefaction and lateral outflow of the basaltic Sima layer of the earth's shell underlying a geosynclinal basin,* together with the granitic root matter of the continental Sial masses and mountain ranges (*see Figures 1.2 and 1.3*). The crustal movements already previously described then follow igneous activity on a similar huge scale; this pours out the equivalent of a 40,000 ft thickness of basaltic and granitic materials as intrusive and extrusive lavas from the lower regions of the Sima beneath the geosyncline, which thus considerably weakens the earth's outer granitic crust by leaving it regionally unsupported in the Sima and Upper Mantle. A shortening of the Sial crust, consequent upon its resettlement into the Sima and a contraction of the interior shell due to its enormous loss of heat, must follow as a natural sequel during the restoration of isostatic equilibrium. All the available evidence points to the fact that we are living today in a quiescent period during such a mountain building epoch which commenced with the Himalayan, Alpine, Rocky and Andean orogenic movements in the Eocene and Miocene periods of the Tertiary Era (*see* Table 1.3), some 40 and 20 million years ago; since then denudation and deposition have continued steadily, while present-day earthquake and volcanic belts follow the lines of these established mountain chains: e.g., along the famous 'Ring of Fire' around the boundaries of the Pacific Ocean on the seaboards of North and South America and China-Japan, via the submerged 'mid-Atlantic Ridge' and the 'Great Rift Valley' of East Africa, extending from the Zambesi river northwards through the Red Sea to the Jordan valley in Israel, together with an east-west axis from the Mediterranean basin to the Himalayas (*see* Table 1.1). There is also the important Brazilian rift zone, the Baikal Rift Zone in the U.S.S.R. and the lesser Rhine Graben, etc., to name a few more tectonically active regions.

While the above explanation suffices for our immediate purposes, there are many problems involved in the study of such large scale crustal movements and the mechanism of their formation. Some authorities contend that

---

* The occurrences may be linked with the theory of Isostacy (Gk. meaning equal poise). According to this theory, the land or continental granitic masses of the earth's crust (the Sial) are in a state of balance resulting from the force of gravity on different thicknesses of crust (which must result in equal pressure acting on equal circumferential areas of the Upper Mantle throughout the whole earth shell), as if they were floating in the denser sub-structure of basaltic material (the Sima) which underlies the continents and oceans alike. Where a mountain chain of lighter density rock forms a large upward projection, there is thus a corresponding bulge beneath into the Sima. Under the oceans the Sial may be very thin, or the rocks of the Sima exposed (which is then much thicker than under the continents, *see* Chap. 1—Mohorovic Discontinuity, Mohole, etc.), as evidenced by rock samples obtained from the floor of the Pacific Ocean, the water of which, together with a much thicker Sima, helps to provide the necessary crustal weight balance per unit of surface area. Any disturbance of this gravitational equilibrium between crustal blocks, by the accumulation of enough sediment in one place, is ultimately righted through the mobility of the earth's outer Sial crust in the Sima (*see Figures 1.2 and 1.3*), and the tendency of isostatic forces to produce equilibrium everywhere. The Scandinavian coastline especially, shows evidence today of gradual uplift since Pleistocene times after removal of the weight of encroaching ice sheets, although this has been a quiet readjustment, not one associated with orogenic activity.

the wrinkling of the earth's surface originates from the steady cooling and contraction of the earth's interior to produce an effect rather like that on the surface of a shrivelled apple. The major movements of the earth's crust are, however, lateral rather than vertical, and are localized in geosynclinal areas, since these approximate to the weak places of the crust at the junction between granitic continental masses and their underlying basaltic layers around ocean boundaries.

Referring to Section 1.8, page 25, the energy released inside rock masses, by the radioactive breakdown of uranium and thorium atomic nuclei, is being continually transformed underground into small amounts of heat which reach the earth's surface to be radiated outwards into space. When the deposition of sediment in a geosynclinal basin, or offshore from a continental mass, blocks this flow, the temperature in the Sima beneath rises to the point when the intermediate solid basalt layer and root rocks of the Sial melt; ultimately the earth's crust becomes weak enough to buckle and vast quantities of this volcanic material are poured out as described in the previous paragraphs.

This subject is very interesting but the student must fill in such further details as he requires by supplementary reading (*see* References and Bibliography, Chaps. 1, 2 and 4).

Summary

It will now be apparent that the 'denudation cycle' will once again recommence on any newly-formed heights, which have a mineral content unstable to the weathering agents operating on their surface. The processes leading to the local or regional deposition of sediment will again continue to the stage where, aeons of time later, a regional mountain building orogenic epoch will emerge from the shattering of the long-term unstable equilibrium of another geosyncline. Between such times local crustal disturbances will, to a minor degree, produce unconformities, cause the folding and faulting of rocks and permit igneous intrusions or extrusions to flow, but the history of deposition is complicated by these and other factors, although the broad picture of the cycle remains valid.

2.7. EXERCISES

1. Describe the mode of occurrence and particular physical features of the following geological formations:
(*a*) River terrace; (*b*) raised beach; (*c*) flood plain; (*d*) glacial overflow channel. (From I.C.E., Pt. II examination, October 1953.)

2. Define the term 'Denudation' and discuss the part that water plays in the disintegration of rock masses. Describe also some of the other processes which lead to the weathering and transportation of rock and summarize

the complete cycle of geological events which includes the denudation and formation of land areas.

3. Describe how ice may modify an existing land surface and indicate some of the problems likely to be encountered in engineering operations in glaciated country. (From I.C.E. Pt. II examination, April 1959.)

4. Write an account of rock weathering, stating the principal weathering agents and describe how they operate. In particular, discuss the processes of marine erosion and their effect and describe briefly the methods adopted to minimize such erosion on coastlines. (From I.C.E. Pt. II examination, April 1958.)

5. With the aid of diagrams describe the following terms:

(a)	terminal moraine	(g)	oxbow lake
(b)	wave-cut platform	(h)	fjord coast
(c)	base-level of erosion	(i)	scree
(d)	hanging valley	(j)	storm beach
(e)	meander migration	(k)	erratic boulder
(f)	glaciated valley	(l)	incised meander

(From I.C.E. Pt. II examination, October, 1955.)

6. What is metamorphism? Sketch and describe any well-known examples of this phenomenon with which you are acquainted.

7. Give a short account of the geological features of a mature river from source to mouth, emphasizing how a knowledge of these features may be of importance in civil engineering operations. (From I.C.E. Pt. II examination, April 1955.)

8. Describe the sedimentary rocks which might be formed in a sea if the land mass providing the sediment is a low-lying granite terrain in the tropics. (From I.C.E. Pt. II examination, October 1958.)

9. Describe the various types of rock which are formed under marine conditions, showing how the character of each rock varies with its distance from the shore-line.

2.8. PRACTICAL WORK—SOME SUGGESTIONS

1. The following picture postcards, photographic booklets, photographic prints, coloured postcards (List No. 3—1963), transparencies (List No. 5—1964), and lantern slides, available for free inspection there or at small cost from the Geological Museum,* South Kensington, booksellers or H.M.S.O. and agents, are particularly recommended as illustrative of the many geological features mentioned in this chapter. An abridged catalogue of 'Classified Geological Photographs' is also obtainable.

* *See* page 40.

Weathering and Denudation

Diorama—The Cheddar Caves (D), Somerset (cave formation: stalactites, stalagmites)　　　　　　　　MNL 515

Diorama—The Cheddar Caves (D), Somerset (cave formation in limestone)　　　　　　　　　　　　MNL 516

Dove Holes, Dovedale, Derbyshire (solution caverns in carboniferous limestone)　　　　　　　　　　CP 20

Rocking Stones, Howden Moors, Yorkshire (atmosphere weathering of millstone grit)　　　　　　　　A 4733

Action of frost in weathering (Aretes and screes), Carn Dearg, Meadhonach, Inverness-shire　　　　　C 1783

Earth pillars, near Fortrose, Ross-shire　　　　　　C 1881

Maviston Sandhills, Moray (forest in process of burial under sand dunes)　　　　　　　　　　　　　B 830

There are also many 'Relevant Museum Exhibits' (RME) and 'Dioramas' (D), for references above and hereafter, available in picture form.

The Great Ice Age and Glacial Phenomena

The Parallel Roads of Glen Roy, Inverness-shire (margins of former glacial lakes)　　　　　　　　　C 2340

Erratic Block, Trefarthen, Anglesey　　　　　　　A 1481

Fluvio-glacial Cone, Glen Fishie, Inverness-shire　　B 758

Glacial Boulder Beds and Gravels, St. Bees, Cumberland　A 2381

U-shaped Valley, Glen Rosa, Isle of Arran　　　　　C 2876

Hanging Valley, Glen Nevis, Inverness-shire　　　　C 1745

Loch Coruisk and Cuillin Hills, Skye (ice-eroded basin)　B 167

Perched Block (Silurian grit on carboniferous limestone), Norber, Austwick, Yorks　　　　　　　　　A 7600

Photographic Booklet No. 3 Glaciated scenery. Snowdon and Cwm Glaslyn (Ordovician rhyolite lavas and volcanic ashes)　　　　　　　　　　　　　　　A 6492

RME—*Edinburgh from Braid Hills* (D), volcanic rock carved by glacial erosion—*see also* CP 10.

An Alaskan Glacier. Eroding action of Dawes Glacier in S.E. Alaska, including moraine formation.

Glaciers and Early Man. Alpine, Himalayan, Alaskan Valley Glaciers. Arctic and Antarctic Ice Formations.

Distribution of Pleistocene Ice Sheet in Great Britain.

Rivers and Waterfalls

Strath Glass, Inverness-shire. Alluvium-filled basin　C 1302

The Avon Gorge (D), Bristol　　　　　　　　　　A 6289

River Wye, Near Fownhope, Herefordshire (Meanders)　A 6250

Tidenham Bend, River Wye, Chepstow, Gloucestershire	A 6276
Cuckmere River, Sussex (Meanders and alluvium)	A 5421
Photographic Booklet No. 4. River action. Scwd-yr-Elra Falls, Ystradfellte, Brecon (D) also	A 4904
Thornton Force, Ingleton, Yorkshire (RME)	A 7625

Coastal Scenery and Marine Erosion

Diorama. Lulworth Cove and Stair Hole (D)	CP 27 ⎫
Dorset. Upper Jurassic and Cretaceous rocks	MNL 513 ⎬
The Needles and Alum Bay, Isle of Wight (D)	MNL 501
High Stacks, Flamborough, Yorks (Stacks and Blowholes in Chalk and Boulder Clay)	A 5443
Selwick's Bay, Flamborough, Yorks (Natural Arch in Chalk)	A 5446
The Seven Sisters, near Seaford, Sussex (Cliffs of Chalk)	A 5414
St. David's, Pembroke (Cliffs of Cambrian sandstone)	A 6089
The Old Man of Hoy, Orkney Islands (Stack of Old Red sandstone)	CA 3312
The Stacks of Duncansby, Caithness (Old Red sandstone)	B 865
Coast Erosion, showing wrecked houses, south of Lowestoft, Suffolk	A 6940
Photographic Booklet, Coastal Views No. 1.	

Scenery of Sedimentary Rocks

High Rocks, near Tunbridge Wells, Sussex (Tunbridge Wells sandstone)	A 5378
Giggleswick Scar, Yorkshire (Millstone grit and carboniferous limestone)	A 7583
Cheddar Gorge, Somerset	A 6038
Dovedale, Derbyshire (Gorge-like valley due to river action in carboniferous limestone)	CP 21

(also RME. Sedimentary rock structures showing false bedding, ripple marks, insolation cracks, animal tracks, etc.)

Each card carries a brief explanatory description which points out the most interesting details of the picture shown. Further detailed references, e.g., The Moine Thrust, photographs etc., are given in other chapters.

Enquiries should be addressed to The Director, Geological Survey and Museum, London, S.W.7.

A full list of maps and publications is available on application, e.g., Regional Handbooks, District and 1 inch sheet memoirs are catalogued in Sectional List No. 45 obtainable from H.M.S.O., Booksellers and Agents.

The Natural History Museum offers a similar service for rocks, minerals and fossils, e.g., *see* N.H.M. Form 22a, List G1–100 postcards, etc.

The United States Geological Survey operates an information service, aspects of which are referred to in Chapter 6. Particular reference should be made to the *Monograph on the Grand Canyon* if readily available.

A large collection of pictures and illustrations may be quickly gathered from magazines and museums additionally to the above sources, and should provide varied examples of geological phenomena.

2. The following list of figures, all except No. 12, Geological sections, are available for study in the Regional Handbooks of the Geological Survey. Many details of fossils relevant to Table 1.3 are also supplied in these books and on postcards.

Pre-Cambrian	General	1. N–S section through Scotland
		2. Canisp, Sutherland
Lower Paleozoic		3. Malvern Hills
Cambrian and Silurian		4. Harlech Dome and also
		4 (a). Clarach Bay Folded Strata
Upper Paleozoic		5. Ledbury—Malvern Hills
		6. South Wales Coalfield
		7. Pennine Uplift area
Mesozoic		8. Midlands to London NW–SE
		9. Weald anticline
Tertiary	Cretaceous to Eocene	10. London Basin and Weald
	Oligocene Miocene	11. Isle of Wight N–S
Quaternary	Pleistocene	12. Pleistocene Ice Sheets
	Recent	13. Lower Thames Valley

The exhibition of 'British Regional Geology'* in the London Geological Museum contains particularly relevant photographs, models, maps, handbooks and memoirs for each of 18 distinct regions.

A special guide to the 'Geological Column' shows the systems and stratigraphy (with fossil examples) of the British Isles.

3. Investigate the locality in which you reside to discover what evidence there is of past changes wrought by geological processes and weathering. Compare the effect of the various weathering agents on local features, man-made and natural; look particularly for landslips. Note: First refer to the appropriate ¼ inch geological map and regional memoir and then continue your study in more detail with the relevant 1 inch map and memoir.

* The geology of the districts around London is separately and specially exhibited, with three particular illustrations and models listed as under:
London and the Thames Valley;
The Wealden District;
The Hampshire Basin and adjoining areas.

2.8. PRACTICAL WORK—SOME SUGGESTIONS

Compare also the topographical O.S. and geological survey maps for identifying land forms and correlate them with the local geology.

4. Examine the chief publications of the Geological Survey of the I.G.S. and familiarize yourself with the means of gaining access to any kind of geological information you may require.

Make careful notes for future reference.

REFERENCES AND BIBLIOGRAPHY

Artz, W. S., von (1952). *An Introduction to Physical Oceanography Reading*. U.S.A.; Addison Wesley

*Bagnold, R. A. (1940–41). 'Beach Formation by Waves'. *J. Instn. civ. Engrs.* **15,** 7; **12** (1939) June

Bagnold, R. A. (1941). *Physics of Blown Land and Desert Dunes*. London; Methuen

Blood, H. K. W. (1949). 'On the Accretion and Erosion of Beaches'. *J. Instn. civ. Engrs.* Paper 5706, No. 6, April

British Standards Institution (1957). *CP2002 Earth Retaining Structures*. App. G. and J. 300 and 400. London

*Conference on North Sea Floods (Jan./Feb., 1963). *Proc. Instn. civ. Engrs.* Dec.

Daly, D. A. (1949). *The Floor of the Ocean*. Oxford University Press

Davies, G. M. (1949). *Students Introduction to Geology*. London; T. Murby

Gamow, G. (1959). *Biography of the Earth*. Revised edn. 1962. London; Macmillan

Garland, G. D. (1966). *Continental Drift*. Oxford University Press

Geological Society (1968). Cloet, R. L., *et al.* 'Marine Studies Group (Sediments)'. *Proc. geol. Soc.* No. 1644, 291

Geological Society (1968). Various authors. Volcanic Studies Group (latest world wide data on volcanism, tectonicism, magmatism, earthquake belts, rift systems, seismic activity, orogenies and geosynclines). *Proc. geol. Soc.* No. 1644, 257

Harbaugh, J. W. and Wahlstedt, W. J. (1967). 'Fortram IV program for mathematical simulation of marine sedimentation with IBM7040 or 7094 computers'. *Kansas State geol. surv. Spec. Dist. Publ.* **9,** 32 pages

Hardy, A. V. and Monkhouse, F. J. (1964). *The Physical Landscape in Pictures*. Cambridge University Press

Holmes, A. (1965). *Principles of Physical Geology*. Revised edn. Edinburgh; Nelson

Horrocks, N. K. (1964). *Physical Geography and Climatology*. London; Longmans Green

Hutchinson, J. N. (1969). Coastal Landslides at Folkestone Warren, Kent. *Géotechnique* **19** No. 1. March, 6–18

Kemp, P. H. (1962). 'A Model Study of the Behaviour of Beaches and Groynes.' *Proc. Instn civ. Engrs.* Rep. No. 6558, **22,** 191

Kuenen, P. H. (1960). *Marine Geology*. New York; Wiley

Lake, P. and Rastall, R. H. (1941). *Textbook of Geology*. 5th Edn. London; Edward Arnold

*Longuet-Higgins, M. S. (1952). 'Mass Transport in Water Waves.' *Phil. Trans. R. Soc.* Ser. A. **245,** 535

Longwell, C. R. and Flint, R. F. (1961). *Introduction to Physical Geology*. London; Chapman and Hall

Nevin, C. M. (1942). *Principles of Structural Geology*. 4th edn. New York; Wiley

Read, H. H. (1949). *Geology*. Oxford University Press

*Report of the Departmental Committee on Coastal Flooding (1954). 'Fieldwork on Stability of Sea Defence Banks in Essex and Kent.' B.R.S., H.M.S.O.

*Reports of the Committee on Coast Protection. *Chart, civ. Engrs.* I.C.E. bulletins May 1965 et seq.

REFERENCES AND BIBLIOGRAPHY

Thornburg, W. D. (1954). *Principles of Geomorphology*. New York; Wiley
Trask, P. D. (1950). *Applied Sedimentation*. New York; Wiley
Twenhofel, W. H. (1961). *Treatise on Sedimentation*, 2 vols. 2nd edn. New York; Dover Publications
*Waters, C. H. (1939). *Equilibrium Slopes of Sea Beaches*. University of California Press
The Wealden District. British Regional Geology London. H.M. Geological Survey
Zumberge, I. H. (1963). *Elements of Geology*, 2nd edn. New York; Wiley

* *See* pages 48 and 67.

CHAPTER THREE

ROCKS AND MINERALS

(Classification—Economic Uses)

3.1. INTRODUCTION

Previously the formation of igneous, sedimentary and metamorphic rocks was discussed in general terms; in so doing, reference was made to some of the most important examples in each family and broad classifications were given. The full geological classification of rocks and minerals, however, is probably too complex and detailed for the engineering student. He should know how to employ simpler classifications for broad groups of rocks or minerals and be able to give, from site explorations, reasonably detailed descriptions; in particular, visual inspection of hand specimens and suitable written reports thereon should form the basis of a student's work.

We must now study the chief types of rocks in regard to their origin, crystalline structure and mineral content, thereby amplifying the scope of any previous remarks. Since minerals are the units from which all rocks are constructed, it will be necessary to describe such minerals as bear directly upon our requisite field of knowledge and this aspect of rock structure, therefore, will be considered first.

3.2. MINERALOGY

The study of the common rock-forming minerals may conveniently begin with a basic outline of the chemical elements and compounds which constitute the majority of such minerals, as listed briefly in Table 1.1 (c); in Table 3.1 a resumé of mineralogy has been given for easy reference.

Although only the relevant rock-forming silicate and non-silicate and ore minerals are mentioned in this text, the references given in this chapter to books on mineralogy should enable the student to refer, when necessary, to any of the 2,000 or so minerals classified by the mineralogist. Mineral names originate generally from Latin or Greek terms, which are related to one or other outstanding physical property of a particular mineral.

Readers familiar with chemistry, will appreciate the chemical formulae given for the various minerals described later in the text, but others may find that to remember the composition in general terms will suffice for their purpose. For example, the composition of any one mineral or compound is constant. Calcite (calcium carbonate—$CaCO_3$) always contains 40 per cent of calcium, 12 per cent of carbon and 48 per cent of oxygen, which facts are embodied most economically in the formula $CaCO_3$.

Examples: Brief outline descriptions of some common and typical minerals follow as a basis for further discussion of mineral properties. They are illustrated in *Figure 3.1*.

Iron pyrites—consist of brassy yellow, metallic, cubic crystals, which give a brownish to black or green streak and break with a conchoidal fracture, although they show no special cleavage. Composed of iron sulphide (FeS_2) and known as Fool's Gold, it has a hardness of over 6, a specific gravity of 5 and is found in many rocks (especially the coal, shale or slate varieties), or *massive* in veins and nodules, but is not worked as a source ore.

Magnetite—an important source ore of iron is easily recognized as black oxide (Fe_3O_4) with metallic cubic crystals which commonly form into octahedra (*Figure 3.1a*); 'magnetite' possesses also the magnetic properties associated with its name and has a poor cleavage with a conchoidal fracture, a hardness of about 6 and a specific gravity of 5.

Haematite—the red oxide of iron (Fe_2O_3) forms metallic grey-black opaque crystals belonging to the hexagonal system, mostly found *massive* in sedimentary beds, or as the 'kidney ore' variety (*Figure 3.5b*) in veins. Similar to magnetite in cleavage, fracture, hardness and specific gravity, it has a distinct red streak and gives an uneven fracture when cleaved.

Dolomite—is a carbonate of magnesium and calcium ($CaCO_3.MgCO_3$), forming pearly white-yellow crystal rhombohedra of the hexagonal system, with curved faces; it is similar to calcite in its cleavage properties, although somewhat harder at the Moh value 3·5–4 and possesses a specific gravity of 2·8. In a minute crystal form, dolomite is the chief constituent of magnesium (or dolomitic) limestone, but it also occurs *massive* with a pale brown colour in ore veins alongside galena and in similar sources to those of fluorspar.

Felspar—(from the German 'Feldspar', meaning 'Rockspar'), is a complex silicate composed of silicon, oxygen, aluminium with either potassium, sodium or calcium which forms the shining, hard, insoluble, oblong pink or white monoclinic and triclinic crystals found abundantly in granite and many other igneous rocks.

Mica—(or Mico-Iglisten) is a complex silicate compound which includes the elements silicon, oxygen, aluminium and potassium, sometimes in combination with magnesium and iron; although from the monoclinic crystal system (*see* Table 3.2) it forms short hexagonal crystalline columns, pearly white (for the muscovite) or black in colour for the biotite variety. Mica is mainly present in granite and other acid igneous rocks, but is also found in numerous sedimentary rocks.

Table 3.1. Mineralogy
(All matter may be divided chemically into two classes: elements and compounds)

1.	Elements	Substances which cannot be split up or decomposed into simpler substances having different properties	
2.	Compounds	Substances containing two or more elements always combined in certain definite proportions	The elements common in rocks, usually in the form of their mineral compounds, are listed in Table 1.1 (c).

3.	Minerals	Are substances:
		 1. Having a definite chemical composition
		 2. Formed naturally by inorganic processes
		 3. Which have a definite internal structure i.e., crystalline
		 4. which have constant physical properties according to the mineral composition and structure
		 The majority of minerals present in *rocks* are chemical compounds such as oxides and silicates; carbonates and sulphates; chlorides, sulphides, etc.
		    The formulae for the composition of all crustal rocks in terms of common oxides are as follows (in percentages):
		 SiO_2 Quartz 59·1* Fe_2O_3 Haematite 3·1*
		 Al_2O_3 Alumina 15·2 (red iron oxide)
		 CaO Lime 5·1 K_2O 3·1
		 Na_2O Soda 3·7 H_2O water 1·3
		 FeO Ferrous 3·7 TiO_2 1·2
		 MgO Magnesium 3·3 P_2O_5 0·3
		           Remainder
		          (*see* Table 1.1 (c))
			A few elements occur uncombined as minerals, e.g., Carbon (C) Sulphur (S) Gold (Au) Copper (Cu)
			
			 *One *atom* of silicon (Si) always combines chemically with two *atoms* of oxygen (O_2) to form one *molecule* of silica or quartz (SiO_2). Similarly, three *atoms* of oxygen always combine with two atoms of iron (Fe) to form one *molecule* of haematite (Fe_2O_3).

4.	Rocks	*Typical examples:* C Graphite Cu Copper Au Gold Silicon dioxide SiO_2; Pure sandstone— Calcium carbonate $CaCO_3$; Pure limestone— Quartz, felspar, mica; Granite—	One element One mineral compound (quartz) One mineral compound (calcite) Two or more mineral compounds.
5.	Mineral recognition tests	*Physical:* 1. Crystalline form 2. Cleavage 3. Hardness 4. Specific gravity 5. Lustre and feel 6. Colour and streak *Optical and Chemical:* 7. Microscopic inspection 8. Blowpipe analysis	All samples of the same mineral have the same physical properties and chemical composition which are invariable. Thus, physical tests suffice for mineral identification

Note: Chemically speaking, an 'atom' is the smallest indivisible particle of any substance and a 'molecule' is the smallest portion of any substance which can have a separate existence; it usually consists of a number of atoms in combination, always in a definite ratio. The molecules of certain elements are *monotomic*, e.g., the solid and metallic elements such as carbon (C), sulphur (S), copper (Cu), etc., but those of gases such as oxygen (O_2), hydrogen (H_2), nitrogen (N_2), etc., are *diatomic* and the 'molecule' of the gaseous 'compound' carbon dioxide (CO_2), for example, consists of 1 atom of carbon and 2 atoms of oxygen in combination. Mention of the most familiar compound, water, molecular formula H_2O, should further illustrate these chemical principles, but if this brief résumé is inadequate for some students, they should consult any elementary chemistry textbook.

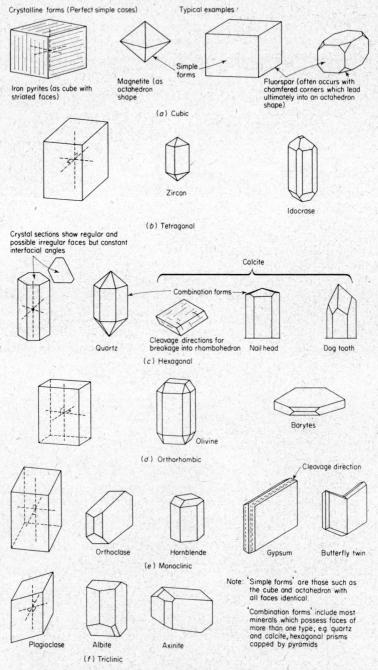

Crystalline forms (Perfect simple cases)   Typical examples

Simple forms

Iron pyrites (as cube with striated faces)

Magnetite (as octahedron shape)

Fluorspar (often occurs with chamfered corners which lead ultimately into an octahedron shape)

(a) Cubic

Zircon

Idocrase

(b) Tetragonal

Crystal sections show regular and possible irregular faces but constant interfacial angles

Calcite

Combination forms

Quartz

Cleavage directions for breakage into rhombohedron

Nail head

Dog tooth

(c) Hexagonal

Olivine

Barytes

(d) Orthorhombic

Cleavage direction

Orthoclase

Hornblende

Gypsum

Butterfly twin

(e) Monoclinic

Note: 'Simple forms' are those such as the cube and octahedron with all faces identical.

'Combination forms' include most minerals which possess faces of more than one type; e.g. quartz and calcite, hexagonal prisms capped by pyramids

Plagioclase

Albite

Axinite

(f) Triclinic

Figure 3.1. Diagrams of crystal systems. (Crystallographic axes indicated by dotted lines)

72

Gypsum—is composed of hydrated calcium sulphate $(CaSO_4.2H_2O)$ occurring as small, soft crystals of monoclinic form, normally white in colour and generally found packed into veins associated with limestone rocks or *massive* in mineral beds similar to those of *rock salt*.

Rock salt—occurs as small cubic crystals of sodium chloride (NaCl) packed into a colourless or white soft mass; it is usually found in beds several feet thick and is easily distinguished by its taste.

Fluorspar—commonly appears as clear, glassy, fairly hard cubic crystals of calcium fluoride (CaF_2), which occur often in veins along with lead and tin ores. It may appear, however, in many colours from yellow through green to blue or purple.

3.3. IDENTIFICATION OF MINERALS

Since the majority of minerals are generally very complex chemical compounds, it is necessary to have some simple physical methods of identification which are both clean and relatively quick. Such recognition or classification tests may be summarized as follows, but it must be remembered that mineralogy is a specialist science of far wider range and content than the succeeding paragraphs indicate (*see* Read, 1953).*

Physical Tests

1. Crystalline form; 2. Cleavage; 3. Hardness; 4. Specific gravity; 5. Lustre and feel; 6. Colour and streak.

Optical and Chemical Tests

7. Microscopic inspection; 8. Blowpipe analysis.

Fortunately, a mineral's basic physical properties are sufficient to provide an identification in many cases and its chemical composition can then invariably be deduced from knowledge of the particular mineral type. (*See* Table 3.1.)

1. *Crystalline form*—In order to apply the first of the tests listed, it is necessary to know the various crystalline forms which occur in nature. This does not mean that every sample inspected automatically possesses flat or curved facial surfaces arranged according to a definite geometrical shape like that of a cube or hexagon, although such a solid outline would show the 'crystal form' completely. The external shape is, however, an indication of the pattern of the internal crystal structure, but the precise external shape developed will depend largely on the availability of space for free crystal growth, which rarely occurs in most natural rock specimens, and masses of very small crystals or fragments of larger ones show no obvious

* *See Rutley's Mineralogy* and other references, page 109.

'crystalline form'. Perfect crystals can be artificially grown and do sometimes occur naturally in the specially favourable circumstances of ore veins or rock cavities (*Figures 3.1* and *3.4*); such specimens exhibit faces which are usually flat but sometimes may be curved, as happens with dolomite. Although crystal form is designated from a perfect external crystalline shape, as given in Table 3.2, it is really the relative positions and lengths of the *x, y, z* axes of the crystal (the crystallographic axes—intersecting in a common origin within the crystal—*Figure 3.1*) which determine its group classification and show the orderly geometric arrangement or lattice of the atomic structure for each particular mineral. The angle between the adjacent faces of any given crystal is, however, constant for each mineral type (e.g., in the *cubic* form—90 degrees, as with fluorspar, or in the *hexagonal* form— 120 degrees, as with quartz) even though the relative size of the different faces may vary appreciably and the true crystalline form is obscured by facial irregularities. The angles between crystal faces are measured by an instrument known as a *goniometer*, of which today there are many refined examples. A draughtsman's ordinary adjustable set square will serve the student as a simple 'contact-goniometer' to be used for checking interfacial angles and their constant value in various specimens of a given mineral.

There are six 'crystal systems', and the brief descriptions of some important minerals, already given, should assist the student to obtain a better understanding of the method of classification by *Form*.

2. *Cleavage*—When placed on a steel plate and struck with a light hammer blow, crystals tend to break or *cleave* along planes parallel to one of the more important faces of the crystal and produce smaller visible fragments of definite shape. This property assists an investigator in identification of *crystalline form* and thence of a rock's constituent minerals. Some minerals like quartz have no particular cleavage planes or directions along which crystals tend to break more readily while others, like mica, cleave easily into very thin flake-like elastic leaves parallel to their basal crystal plane; calcite, for example, has three distinct cleavage planes inclined to each other, along which its crystals break into *rhombohedra* (*see Figure 3.1c*).

The various *degrees* of cleavage are called 'Perfect', 'Good', 'Distinct', 'Imperfect', according to their quality; mineral cleavage is also important in considering rock architecture and the use of building stones generally, affording as it does, definite shearing planes in a rock for easier quarrying and dressing purposes of the extracted blocks into suitably sized stones.

3. *Hardness*—If one substance is harder than another, the former will scratch the latter. On this basis, a Table of Hardness can be drawn up in which each member substance scratches the one numbered below it in the table and is itself, in turn, scratched by those numbered above it. This characteristic property gives one of the best means of identification adopted

for minerals and is thus of great value to the investigator; the precise scale for scientific comparison of mineral hardness, called *Mohs' Scale*, is formed from 10 standard minerals as listed in Table 3.3, in ascending numerical order from the softest mineral at No. 1.

Table 3.2. The six crystal systems
(*See Figure* 3.1)

System	Remarks	Examples
(a) Cubic	(or Isometric), includes octahedron, dodecahedron and compound crystals. Three equal axes at right angles with nine planes of symmetry*	Fluospar, magnetite, garnet, rock salt, leucite, pyrite and zinc blende
(b) Tetragonal	Two equal lateral axes making angles of 90 degrees with each other and a vertical axis, longer or shorter than the other two, giving five planes of symmetry	Zircon, idocrase, cassiterite (tin stone)
(c) Hexagonal	Three equal lateral axes making angles of 120 degrees with each other and a vertical axis, longer or shorter than the other three, giving seven planes of symmetry	Quartz, calcite, apatite, tourmaline, beryl, nephaline
(d) Orthorhombic	(cubic system distorted horizontally) Three unequal axes, all at right angles. Three planes of symmetry	Olivine, barytes, topaz, enstatite
(e) Monoclinic	(a distortion of orthorhombic) Three unequal axes, one vertical, one at right angles to the vertical axis, the third making an angle with the plane containing the other two. One plane of symmetry only	Horneblende, gypsum, augite, orthoclase felspar, mica
(f) Triclinic	(a distortion of monoclinic) Three unequal axes, none at right angles. No planes of symmetry	Plagioclase felspars, axinite
(g) Amorphous	A material having no crystalline form	

Special descriptive terms

The characteristic mass shape of a crystalline block is termed Eumorphic.

A 'pseudomorph' is a solid material which takes on the crystalline form of some other substance, e.g., hollow cubes of mud which have the form of rock salt crystals are found in the spaces left behind by dissolved salt crystals under water and are termed 'salt pseudomorphs'.

---

* A 'plane of symmetry' divides any crystal into two exactly similar pieces, each a mirrored image of the other. Minerals of the simplest composition possess the simpler forms such as 'cubic', 'tetragonal' (a, b, c), whereas those with a more complicated molecular structure are found in the distorted forms 'orthorhombic', monoclinic' and 'triclinic' (d, e, f).

Thus the hardness of the finger nail is approximately 2·5–3 on the Mohs' scale and that of a pocket knife is about 6.

Table 3.3. Mohs' scale of hardness

1. Talc	may be scratched with the finger nail	6. Orthoclase felspar	
2. Gypsum		7. Quartz	
3. Calcite	may be scratched with a pocket knife	8. Topaz*	Cannot be scratched with a pocket knife
4. Fluorspar		9. Corundum*	
5. Apatite		10. Diamond*	

* Gem stones.

Hardness is a very dependable quality, for any piece of the same mineral, irrespective of origin, shows the same value. The test procedure is to scratch an unknown mineral with the known minerals of the Mohs' scale until it can be placed in the numbered scale between two of the standard minerals. With a finger nail (hardness No. 2) and a pocket knife (hardness No. 5) three ranges of mineral hardness can be distinguished and a sample may be placed approximately near its correct value in Table 3.3. A mineral harder than *felspar* (e.g., felspar is scratched by it) but softer than *quartz* (e.g., it is scratched by quartz) would have a hardness of 6·5. With a finger nail and pocket knife only it is possible to distinguish between *quartz* (No. 7) and *calcite* (No. 3) by means of their 'Mohs' values, although their crystalline appearance may be very similar.

4. *Specific gravity*—The specific gravity (sp. gr. or relative weight) of different minerals is also a most useful guide to classification; specific gravity is by definition, *the weight of a given volume of any mineral compared to the weight of an equal volume of water*. The specific gravity values of the common minerals range from a figure as low as 1 (i.e., the same density as water) to over 20. For instance, the rare metal platinum has a specific gravity of 21·46, but most minerals possess a value lying between 2 and 7 and these figures in the M.K.S. and C.G.S. systems are the same numerically as the density in grammes per millilitre. Thus the density of quartz is 2·65 g/ml³ (and its sp. gr. is 2·65) while that of olivine is 3·5 g/ml³, although the hardnesses of these two minerals are practically equal, being 7 and 6·5 respectively. (Table 3.4.)

Textbooks on mineralogy describe the apparatus used and give details of standard tests for specific gravity determination as well as other suitable methods for determining the specific gravity or density of minerals, which will not be described here (*see* Read, 1953).*

5. (*a*) *Lustre*—is the appearance of a mineral surface given by reflected light, the precise degree of *lustre* depending on the amount of reflection that takes place at the surface of the mineral. Various terms used to describe

* *See also* References page 109.

76

differing degrees of lustre are: Metallic—as galena; Silvery—as graphite; Brassy—as pyrites; Glassy—as quartz; Pearly—as mica or talc; Resinous or Silky—as opal (also called Opalescent). Where there is *no lustre* the mineral is designated as 'dull'.

(*b*) *Feel*—is likewise often a distinctive mineral property, as for example, the greasy or soapy feel of serpentine or talc.

6. *Colour*—(*a*) *In the mass* the colour of minerals can be very variable and is not to be relied upon with any certainty as a guide to mineral recognition; e.g., chlorite is generally green, as is olivine, but other minerals such as quartz, which is normally colourless, may take on the colour of any minor chemical impurities.

(*b*) More reliable as an identification test is the colour of a mineral in its powdered condition, called the *streak* and produced by crushing or rubbing a corner of the mineral crystal on a flat piece of white unglazed porcelain, since this crystal property is constant despite any variations in *mass* colouring. Thus: Haematite (ferrous oxide Fe_2O_3), while grey or black in the mass, gives a *red streak;* Limonite $\left(hydrated-\left\{\frac{Fe_2O_3}{nH_2O}\right\}\right)$, though amorphous in the mass and brown to yellow in colour, gives a *brown streak;* Magnetite (ferric oxide Fe_3O_4), black in the mass, gives a *grey streak;* Iron pyrites (iron sulphide FeS_2), brassy yellow in the mass, gives a *black powder streak*.

NOTE: In the Hardness Test mentioned above, it is important to verify that a scratch really has been made on the softer mineral and that which appears as a scratch is not just a surface 'streak' of the latter in its powder form rubbing off on the harder mineral.

The student should realize that there are other mineral characteristics which are helpful in analysis, such as their optical properties.

7. *Optical properties*—For each mineral these may be summarized as:

(*a*) The effect of a crystal in its *refraction* of light and its value of Refractive Index on different internal planes.

(*b*) A crystal's effect on *polarized* light and the use of polarization colours for identification of the crystallographic axes and/or chemical composition.

(*c*) The detailed textural appearance under a Petrological microscope, especially when a rock specimen has been cut into thin sections.* (*See Figures 3.2* and *3.3*.)

To the mineralogist these are some of the most important and decisive analytical tests used for classifying minerals, and for further information,

---

* Special exhibits in the London Geological Museum of the I.G.S. show the preparation of thin slices of rock for detailed microscopic examination; samples of the principal rock types thus prepared are illustrated in multi-coloured reproductions; also displayed is the sectional appearance of rocks under polarized light with the optical properties of their constituent minerals explained. (*See* Chap. 2, pages 40, 65 and Chap. 3, Practical Work, page 108.)

Table 3.4. Primary minerals

Group or family	Mineral type	Generalized chemical composition	Crystalline form	Lustre
Silicate minerals	*Olivine*	*Silicate of iron and magnesium*	*Orthorhombic*	Vitreous
Pyroxenes (good prismatic cleavage at 90 degrees—Common in the more basic rocks)	Enstatite	Complex silicates of magnesium	*Orthorhombic*	Vitreous to metallic
	Hypersthene	Magnesium and iron	*Orthorhombic*	Sub-metallic
	Augite	Magnesium aluminium and calcium	*Monoclinic*	Vitreous to resinous
	Aegirine	Silicate of iron and sodium	*Monoclinic*	Vitreous to resinous
Amphiboles (cleavage at 120 degrees—Common in the more acid rocks)	*Hornblende*	Complex alumina silicate	*Monoclinic*	Vitreous to resinous
	Tremolite	Silicate of calcium and magnesium		Vitreous
	Actinolite	Silicate of calcium and magnesium iron		Vitreous
	Asbestos	Fibrous form of actinolite		
Micas (Perfect basal cleavage into thin sheets)	*Muscovite*	Silicate of alumina and potassium	*Monoclinic*	Pearly
	Biotite	Silicate of alumina and potassium and iron	*Monoclinic*	Pearly
	Phlogodite	Silicate of alumina and potassium and magnesium	*Monoclinic*	Pearly
Felspars Most abundant of all silicate minerals —essential constituents of igneous rocks	Microcline	Silicates of alumina and and potassium	*Monoclinic*	Vitreous to pearly
	Orthoclase (Straight cleavages at 90 degrees)	Silicates of alumina and potassium	*Triclinic*	Vitreous
Plagioclase Felspars (skew cleavages at 86 degrees)	Albite	Silicates of alumina and sodium	*Triclinic*	Vitreous
	Oligoclase		*Triclinic*	Resinous
	Andesine		*Triclinic*	Sub-vitreous
	Labradorite		*Triclinic*	Sub-vitreous
	Anorthite	Silicate of alumina and calcium	*Triclinic* (lamellar twinning)	Vitreous
Silica (no cleavage)	Quartz	Silicon dioxide (chemically an oxide and structually a silicate)	*Hexagonal*	Vitreous
Accessory minerals (Found in small amounts—do not affect naming of rock)	Tourmaline	Boro-silicate of aluminium	*Hexagonal*	Vitreous
	Garnet	Silicate of Ca, Mg, Fe, Mn, Al	*Cubic*	Vitreous
	Zircon	Silicate of zirconium	*Tetragonal*	Adamantine

NOTE: Minerals in italic are the most common examples of each crystalline form. *See also* Table 1.1 (c)

3.3. IDENTIFICATION OF MINERALS

Table 3.4. Primary minerals

Mass colour	Streak	Fracture	Hardness	Sp. gr.	Common occurrence as a main constituent of igneous/metamorphic rock family named
Olive green	—	Conchoidal	6·5	3·5	Basic and ultrabasic rocks (Peridotite) Olivine basalt, gabbro
Dark brown and green	—	—	5·5	3·2	Basic and ultrabasic rocks (Peridotite) and Intermediate
Dark brown and green	—	Uneven	5–6	3·5	Basic rocks (norite) gabbro
Black	White and grey	—	5–6	3·2–3·5	Basic rocks (gabbro), dolerite, basalt
Brown	—	—	6	3·5	Soda rich igneous rocks (nepheline-sucnite) basalt
Black	Pale green	Uneven	5–6	3–3·4	Acid and intermediate rocks (granite, hornblende, andesite, diorite, syenite)
White	—	—	5–6	3–3·4	Metamorphic rocks
Green	—	—	5–6	3–3·4	Metamorphic rocks
White, brown 'puff pastry'	—	—	2–2·5	2·85	Acid igneous (granite, muscovite) rocks
Black and dark green	—	—	2·5–3	3	Acid igneous (granite; Brotite, andesite) rocks
Colourless, white	—	—	2·5–3	2·75	Metamorphosed limestone
White, pink and red	White	Conchoidal and uneven	6	2·57	Acid igneous (granite) rocks
Greyish white	—	—	6–6·5	2·55	Acid igneous (granite) rocks
Greyish white	—	Uneven	6–6·5	2·63	Acid and intermediate rocks
Greyish white	—	Conchoidal	6–6·5	2·65	Intermediate rocks
Greyish white	—	—	5–6	2·69	Intermediate rocks
Grey	—	Conchoidal and uneven	6	2·67–2·76	Basic rocks
Colourless or white	—	Conchoidal	6–6·5	2·76	Basic rocks
Colouress when pure	—	Conchoidal	7	2·65	Acid rocks Granites Quartz diorite, dolerite
Black	—	Uneven	7–7·5	2·94–3·2	Acid rocks
Dark red and brown	—	—	6·5–7·5	3·5–4·2	Metamorphic rocks
Yellow or colourless	—	Conchoidal	7·5	4·7	Acid rocks

interested students should consult a suitable reference book (*see* Read, 1953).*

The foregoing are essentially special instrumental tests but other simple physical properties are those of *transparency*, i.e., whether a crystal is *opaque*, *translucent*, or *transparent* to light rays; in particular, the crystal property called *translucency* may be ascribed to any crystal having a similar appearance to that of light when seen through frosted glass.

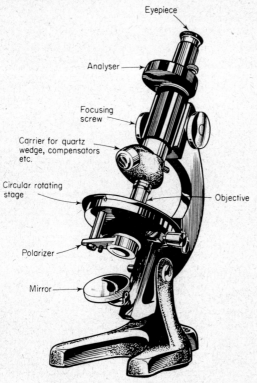

Eyepiece

Analyser

Focusing screw

Carrier for quartz wedge, compensators etc.

Circular rotating stage

Objective

Polarizer

Mirror

Figure 3.2. The prospectus petrological microscope (Note the analyser and Bertrand are enclosed in dust-free cell below eyepiece)

The appearance of the *fracture*, as distinct from the cleavage, of a broken crystal surface is described as:

(*a*) Conchoidal (i.e., shell-like) or curved, such as occurs with quartz or flint;

(*b*) Even—when a fractured surface is smooth.

(*c*) Uneven—when a fractured surface is rough.

(*d*) Hackly—when the fractured surface has small, sharp irregularities.

* *See also* references page 109.

The *form* and *character* of minerals or their crystals and crystal faces when and as they occur naturally in a mass state; e.g., granular, fibrous, scaly, etc.

Form—is designated amorphous* (or non-crystalline), finely crystalline, massive, or eumorphic* and is not to be confused with the property of 'crystalline form' already described in Section 3.3, page 73. It is a help also to note the character of a crystal's faces; for example, the faces of a quartz crystal often have distinctive striations (or lines) which make them readily distinguishable from those of calcite.

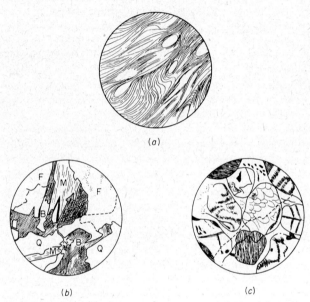

(a)

(b)   (c)

Figure 3.3. Crystal sections. (a) Thin section of obsidian, glassy fine-grained volcanic lava showing flow lines around quartz crystals. (b) Thin section of ordinary 2 mica granite (× 20) showing quartz, Q, felspar, F, biotite, B, and muscovite, M. (c) Thin section of quartzite (× 40) with rounded quartz grains set in a solid quartz matrix

((b) and (c) reduced one half on reproduction)

8. *Chemical analysis*—The simple blow-pipe method of testing by flame colour is very useful in the field (*see* Read, 1953)†. A complete chemical analysis is necessary to discover the full composition of any mineral and the results obtained are, of course, absolute, leaving no margin for speculation which might happen when some of the simple physical tests only are adopted. One easy chemical test is the use of a drop of dilute hydrochloric acid on a mineral specimen. Quartz (silicon oxide) for example, shows no

* *See* Table 3.2.
† *See also* References page 109.

reaction, being chemically inert to all but the very strongest acids, but calcite (calcium carbonate) effervesces and gives off carbon dioxide gas. Calcite and similar minerals are attacked by most acids, quartz by almost none, however concentrated.

As mentioned previously, the number of common minerals of direct interest to the student engineer is comparatively small, a minimum of about 50 altogether, and these types are to be found in the two broad classifications, *primary* and *secondary*. Of all mineral types, the silicate families are the most prolific and distinctive in properties according to the chemical percentage of acidic silicon dioxide (SiO_2) present in a particular mineral compound and whether this represents an excess of silicon and oxygen.

3.4. THE PRIMARY MINERALS

The primary (or silicate) minerals are formed in molten igneous rock magma rising from the depths of the earth at high temperatures and pressures, as indicated in Chapter 1. The more stable minerals are the harder examples like quartz or felspar. A summary of the main types and their properties is given in Table 3.4, and descriptions of the commoner examples in order 'acid to basic' should be attempted by the student from the information given in Table 3.4 and as the following paragraphs indicate.

Quartz is the outstanding example of a primary mineral, occurring as a 'rock former' in granite or other acid to intermediate igneous types and sandstones, etc. Its crystals take the form of either light hexagonal prisms capped by hexagonal pyramids (*Figure 3.1c*) or are just two pyramids arranged with their bases end on end, often with their faces marked by horizontal striations which makes their identification much easier. (*Figure 3.4.*)*

Although chemically 'quartz' is an oxide, structurally it belongs to the silicate mineral family and is ordinarily a colourless 'rock crystal' of pure silica (SiO_2); but even minor impurities make the mass colour test an uncertain guide. For example, the variety known as amethyst is violet coloured; rose quartz is pink; smoky quartz is brown, and milky quartz is white.

The crystals possess a translucent, glassy lustre and a hardness of No. 7— i.e., greater than that of a knife; quartz minerals also have a specific gravity of 2·65 and no particular cleavage planes, while any broken surface gives a smooth conchoidal fracture.

---

* See *Rutley's Mineralogy* and other references, page 109, for an account of the complex physico-chemical processes involved in crystallization of either a single substance like quartz, or a molten magma passing through metastable and labile conditions. Such work, although outside the scope of this present study, concerns not only the petrologist, but also the chemist, metallurgist and others concerned with basic manufacturing industries—iron and steel, alloys, glass, cement, etc.

Hornblende—occurs in igneous and metamorphic rocks, e.g., diorites, andesites and hornblende schist. It is a silicate of calcium, magnesium and iron, black or dark green in colour, and consists of six-sided monoclinic crystals (*Figure 3.1e*), transparent in thin section, with two cleavages at 120 degrees. These show a definite uneven fracture and the crystals are normally opaque with a hardness value of 5–6 and a specific gravity of 3·2–3·5.

Augite—is much like hornblende in appearance and occurs in many basic igneous rocks (e.g., gabbro).

In general, the iron magnesium silicate family consist of the dark, heavy, fairly hard basic minerals, while the acid silicate family form the majority of the light coloured minerals.

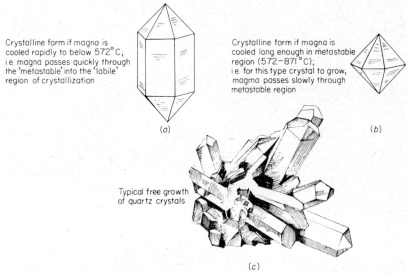

Crystalline form if magna is cooled rapidly to below 572°C; i.e. magna passes quickly through the 'metastable' into the 'labile' region of crystallization

Crystalline form if magna is cooled long enough in metastable region (572–871°C); i.e. for this type crystal to grow, magma passes slowly through metastable region

(a)

(b)

Typical free growth of quartz crystals

(c)

Figure 3.4. Crystalline quartz

3.5. THE SECONDARY MINERALS

The secondary minerals are produced by disintegration and recombination of the chemical constituents of the primary types, usually in acidulated water containing carbon dioxide (CO_2) or sulphur dioxide (SO_2), dissolved from the atmosphere. Most stable, inert and hard minerals like quartz remain unaffected by natural processes, although crystals may be fragmented and their grains worn down. Dissolved potassium (K) and sodium (Na) compounds are carried down to the sea by rivers and calcium (Ca), in combination with carbon dioxide (CO_2) from the atmosphere, forms calcium carbonates $(CaCO_3)$ some of which may become the mineral 'calcite'.

A summary of the main types and their properties is given in Table 3.5.

Table 3.5. Secondary minerals

Group or family	Mineral type	General chemical composition	Crystalline form	Lustre
Secondary by change of pre-existing minerals due to water action	Chlorite	Hydrated silicate of Al, Fe and Mg	Monoclinic	Pearly
	Talc	Hydrated silicate of Mg	Monoclinic	Pearly
	Serpentine	Hydrated silicate of Mg	Monoclinic	Sub-resinous
	Kaolin	Hydrated silicate of alumina	Monoclinic	Dull and earthy
	Epidotite	Si of Al with water	Monoclinic	Vitreous
Silicates	Zeolites: (1) Analcite	Hydrated silicate of sodium	Cubic	Vitreous
	(2) Natrolite	Hydrated silicate of sodium and Al	Orthorhombic	Vitreous
	Chalcedony	Cryptocrystalline* silica	Orthorhombic Really a rock	Waxy
	Flint	Cryptocrystalline* silica	—	—
	Chert (Impure flint)		—	—
	Opal	Hydrated silica	Amorphous	Pearly
Non-silicates Oxides— metallic or minerals	Magnetite	Oxide of iron	Cubic	Metallic
	Haematite	Oxide of iron	Hexagonal	Metallic
	Limonite	Hydrated iron	Amorphous	Sub-metallic
	Ilmenite	Oxide of iron and titanium	Hexagonal	Sub-metallic
	Cassiterite	Tin oxide	Tetragonal	Adamantine
Carbonates	Calcite	Carbonate of lime	Hexagonal	Vitreous
	Dolomite	Carbonate of Mg and Ca	Hexagonal	Vitreous
	Siderite	Carbonate of iron	Hexagonal	Vitreous or pearly
Phosphate	Apatite	Phosphate of calcium	Hexagonal	Vitreous
Sulphates	Gypsum	Calcium sulphate	Monoclinic	Pearly
	Barytes	Barium sulphate	Orthorhombic	Vitreous
	Pyrite	Sulphide of iron	Cubic	Metallic
Sulphides and metal ores	Marcasite	Sulphide of iron	Orthorhombic	Metallic
	Pyrrhotite	Sulphide of iron	Hexagonal	Metallic
	Galena	Sulphide of lead	Cubic	Metallic
	Zinc-Blende	Sulphide of zinc	Cubic	Resinous
Fluoride	Fluospar	Calcium fluoride	Cubic	Vitreous
Chloride	Rock-salt	Sodium chloride	Cubic	Vitreous

NOTE: All minerals in italic type are the more common examples of each crystalline form. No triclinic crystals are listed and some substances like opal (a semi-precious gemstone) are amorphous while flints, etc., are really a cryptocrystalline rock.
* Textural terms are cited in Table 3.6.

3.5. THE SECONDARY MINERALS

Table 3.5. Secondary minerals

Colour	Streak	Fracture	Hard-ness	Specific gravity	Occurrence as common constituents of rock named
Green	—	—	1·5–2·5	2·6–3·0	Metamorphic types
White	—	—	1	2·7	—
Green-black	—	Conchoidal	3–4	2·5–2·6	Basic and ultra-basic types
White	—	—	2–2·5	2·6	Weathering of felspars to form clays
Green	—	Uneven	6–7	3·25–3·5	Metamorphic types
White	White	Subconchoidal	5–5·5	2·25	Basic lava gases
White	White	—	5–5·5	2·25	Basalts
White-grey	—	—	—	—	—
Black	—	Conchoidal	—	—	Upper chalk
—	—	Flat	—	—	—
White-grey	—	Conchoidal	6	2·2	Cracks in igneous rock
Black	Black	Subconchoidal	5·5–6·5	4·9–5·2	Small amounts in igneous rocks
Black or steel grey	Red	Subconchoidal	5·5–6·5	5	Ore pocket, replacing limestone
Yellow	Yellow	—	5–5·5	3·6–4	—
Black	Black	Conchoidal	5–6	4·5–5	Basic igneous rocks
Black or brown	White	Subconchoidal	6–7	6·4–7·1	Ore veins and placer deposits
White yellow	White	Conchoidal	3	2·7	Limestones
White yellow	—	Conchoidal	3·5–4	2·8	Limestones
Pale yellow	White	Uneven	3·5–4·5	3·7–3·9	Limestones
Pale green (varies)	White	Conchoidal	5	3·2	Accessory mineral in igneous rocks
Colourless or white	—	—	1·5–2	2·3	Saline residue
Colourless or white	White	—	2·5–3·5	4·6	Vein deposits
Bronze yellow	Green	Conchoidal	6–6·5	4·8	Ore veins and accessory rocks
Pale yellow	—	—	6	4·9	Sedimentary types
Reddish bronze	Dark grey	Uneven	3·5–4·5	4·4–4·65	Basic igneous
Grey	Grey	Flat	2·5	7·2	Ore veins
Black	White	Conchoidal	3·5–4	3·9–4·2	Ore veins
Green and purple	White	Conchoidal	4	3·25	Ore veins
Colourless and white	—	—	2–2·5	2·2	Saline residues

85

Illustrative descriptions of two important examples, based on data abstracted from Table 3.5, follow as an indication of how to utilize the facts given.

Calcite—is a white glassy crystalline calcium carbonate ($CaCO_3$) common to the limestone rocks. The crystals form hexagonal prisms with either near-flat ends or end-on pyramids and exhibit three perfect cleavages at 60 degrees, causing their breakage into regular 'rhombohedra' with smooth faces of parallelogram shape; Nail head and Dog tooth spars are two famous varieties of these crystal types (*Figure 3.1c*). Calcite crystals are just harder than a finger nail, with a Moh's value of 3 and sp. gr. 2·7 and a tendency to mass coloration by impurities. It is because calcite possesses good cleavage that the crystals break into *rhombs* and these into yet smaller *rhombs*, all with a vitreous lustre; thus, although the fracture is conchoidal, this cleavage property is distinctive in comparison with quartz for example, as are also the results of a simple chemical acid test and that for hardness value.

Calcite is a major constituent of limestones in the form of minute crystals, which also make up the characteristic 'Stalactites' and 'Stalagmites' of underground caves (*see* Chap. 2), but it frequently occurs also as large crystals in veins (*Figure 3.5*).

Galena—the sulphide and most important ore of lead (PbS), possesses distinctive heavy and dull metallic grey (i.e., 'leaden') crystals of the cubic system, which cleave easily into smaller cubes giving an uneven fracture. These cubes have a hardness of 2·5 and a sp. gr. 7·5, but if heated on charcoal in a blowpipe flame, globules of metallic lead easily form. Galena usually contains some silver sulphide and occurs in veins associated with zinc blende, quartz and calcite found in a wide variety of rocks, as with the limestones of Derbyshire and at Broken Hill, N.S.W., Australia.

3.6. IGNEOUS ROCKS

Introduction

Almost three-fifths of the earth's crust consists of silicon dioxide (SiO_2) in one form or another and this acid compound forms silicates of aluminium, iron, calcium, potassium, magnesium and soda (the primary minerals); igneous rocks are aggregates of these primary silicate minerals; but the acid (i.e., excess) silica content, when allied to their 'mode of origin' (*see* Section 1.6), is the most convenient index for chemical classification of various different rock types into groups.

Basis of Classification by Mode of Origin and Chemical Composition

In an 'acid' rock such as granite (often termed, *oversaturated**), the original silica content was in excess of chemical compound requirements for the alkalis or *bases* present in the cooling magma, and that remaining *free* (i.e., if more than 10 per cent of the total silicon dioxide (SiO_2) content) crystallized out as the mineral quartz (*free* silica), thus forming a 'matrix' in which the other primary mineral constituents became embedded. Acid rock is light-coloured, light in weight (sp. gr. 2·7), with mineral quantities of quartz, orthoclase felspar (the silicate of aluminium and potassium), and mica (the silicate of aluminium and magnesium).

In 'basic' rocks such as gabbro (termed *undersaturated**), free silica is deficient in quantity or entirely absent; thus the dark coloured ferro-magnesian silicates like plagioclase (soda-lime) felspar and augite, or olivine and hornblende (basic magnesium silicates) predominate, because the original magmatic silica content was inadequate for complete chemical combination with the iron and magnesium bases present. Oxides of iron (e.g., magnetite and ilmenite) may also be present, especially if the original magma was rich in iron, and the specific gravity of these rocks is greater than 3.

Intermediate rocks (termed *saturated**), generally possess some mineral quartz, with a predominance of ferro-magnesian silicate 'hornblende' and the 'plagio' felspar minerals. As their name 'intermediate' suggests, they lie approximately midway in the possible range of silica content between 'acid' and 'basic' types. Here also a brief mention should be made of the 'ultra-basic' rock types, such as serpentine, which contains no quartz and virtually no felspars; they are very dark indeed, being composed of augite and olivine only, and are denser than all other types with sp. gr. > 3·3.

A most convenient tabular form of classification can therefore be built up from nine distinct groups of igneous rocks, based on the silica content as one direction and their grain size as the other (i.e., their chemical composition and mode of origin or 'crystalline texture') are as shown in Table 3.6.

Intermediate Groups

These rocks are particularly distinguishable under the petrological microscope by their characteristic felspars with interstitual quartz crystals showing; the presence of normal alkali felspars indicates a rock of the Syenite/Trachyte family (*a*), while calc-alkali felspars are indicative of the more basic rocks belonging to the Diorite/Andesite family (*b*), marked in Table 3.6.

---

* The term 'Basic' rock is derived from the richness of the chemical bases present in these types. The various silicate mineral compounds are composed of silica in chemical combinations with the basic oxides of Na, K, Al, Fe, Mg, etc., to form silicate salts like felspar. *Oversaturated, saturated* and *undersaturated* refer to the acid-base chemical proportions of the constituent rock minerals in igneous types and especially the presence or absence of the mineral quartz.

quartz (SiO_2) is thus usually the 'matrix' mineral in an acidic granite rock, being the lightest and last to crystallize out of solution, and hence filling the irregular spaces between the other earlier crystallizing 'accessory, basic, intermediate', felspar and mica minerals; this order of successive mineral precipitation follows from the silica content of each class of mineral, e.g., 'accessory' + basic—minimum silica—heaviest, to quartz, maximum silica—lightest. Also this phenomenon has particular relevance to the formation of slags in metal ore refining, the use of furnace linings and the manufacture of cements, etc. (*see* Table 3.4 for sp.gr. comparisons).

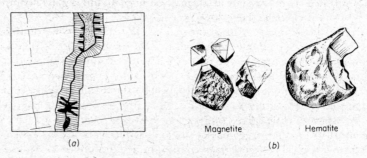

Magnetite   Hematite

(a)   (b)

Figure 3.5. (a) *Calcite crystals grown in solution channel—limestone ore vein minerals shown black.* (b) *Magnetite and haematite*

Plutonic

Granite and granodiorite consists of the light-coloured coarse-grained minerals, whitish quartz, pink orthoclase felspar and black or white mica together with some white or grey plagioclase felspar which latter increases in quantity relative to the other minerals in 'granodiorite' rock types. Granites contain a total of 66 per cent or more acidic silicates and can often be split along definite joint planes; the preparation of the road setts and kerbstones is particularly facilitated when sets of jointing exist in the rock in perpendicular directions to each other.

Diorite is practically a granite type rock with very little quartz and containing more of the 'plagioclase' (soda-lime type) than 'orthoclase' felspar minerals, and with the ferro-magnesian silicate *hornblende* as a major constituent (i.e., with 66–52 per cent of acidic silicates in total).

Dolerite and *Gabbro* rocks largely contain the dark-coloured basic augites and olivine or magnetite minerals in approximately equal quantities, together with a good proportion of plagioclase felspars (i.e., 52–45 per cent acidic silicates in total). Dolerite is more an intermediate/hypabyssal type rock and usually has very small crystals compared to gabbro.

The texture of granite rocks is noted in Table 3.6 which also gives details of

90

the typical chemical composition for an ordinary 2-mica type granite* (*Figure 3.3b*); the *Mineral Composition Chart* shows this mineralogical composition of granite diagrammatically, as it does also for the diorite-syenite and gabbro rock families.

Volcanic

Rhyolite is a light-coloured acid lava with very fine white and pink crystals of quartz and orthoclase felspar; sometimes odd crystals are large, giving the rock a porphyritic texture (*Figure 3.6*), but it is mostly fine-grained. Rhyolite lavas tend to be *vesicular*, thus forming a pumice, but some lava may also form the glasslike *obsidian* (dark bottle).

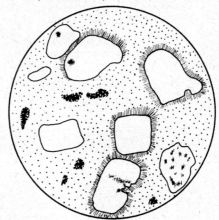

Figure 3.6. Quartz porphyry in thin section (× 20). Note the 'phenocrysts' large crystals, in a fine even-grained matrix (Table 3.6, see Textural Terms)

(Reduced one half on reproduction)

Trachyte is very finely crystalline, light grey in colour, with a parallel-flowing and even grain, but again often porphyritic in texture; plagioclase and orthoclase felspars predominate with a little quartz, while some basic mica and hornblende minerals are also present.

Andesite is similar to trachyte in texture and composition, but of a considerably darker colour, as it tends to contain more of the ferromagnesian mineral types such as mica, hornblende and augite.

Basalt is a very dark fine-grained rock, often nearly black, and closely related in mineral composition to gabbro. It frequently appears in the form of hexagonal columns with 'columnar jointing', which occurs in widespread lava flows. (*See Figure 4.19.*)

Hypabyssal

The three 'Micro' types are simply finer-grained versions of their plutonic

* Various types of *granite* are named by their most distinctive mineral constituent: hence,

Biotite
Muscovite ⎱ *granite*. There are also many varieties of other rock types: hence,
Hornblende ⎰
Nephelin ⎱ basalt, Horneblende syenite, Quartz diorite, Olivine gabbro
Olivine ⎰

equivalents described previously; hence they are named Micro-granite, Micro-diorite and Micro-gabbro in order of chemical composition, 'acid', 'intermediate' and 'basic'.

Igneous rock types are thus readily grouped according to their crystal sizes (or texture) and silica (i.e., quartz, etc.) mineral content into one of the nine main sections listed. Their individual silicon dioxide content varies, Acid, Intermediate to Basic from approximately 75–45 per cent of the total chemical content for each of the three main textural groups as exemplified by the order of the rock examples described previously.

Remember also that the acidic silicon dioxide compounds in conjunction with water form silicates from the basic oxides of Al, Fe, Ca, Mg, potash and soda, etc. and the many different primary (and secondary) silicate minerals account for the wide variation in all rock types, igneous, sedimentary and metamorphic.

3.7. SEDIMENTARY ROCKS

Sedimentary rocks, which were described briefly in Sections 1.6 and 2.2, according to their mode of formation (Table 1.3), and which form the greatest part in area of the visible mass of the earth's crust (namely three-quarters of the rocks exposed on the various land surfaces), are extremely difficult to classify logically; they may, however, be separated into six

Table 3.7. Brief classification of sedimentary rocks

Mainly fragmental or mechanical		Mainly Chemical and Organic			
(a) Arenaceous	(b) Argillaceous	(c) Calcareous	(b) Carbonaceous	(e) Siliceous	(f) Precipitaceous
Boulders	Clays— fireclays	Chalks	Peats	Diatomaceous earths	Salt beds
Gravels	China clays	Limestones	Lignites	Flints	Gypsum beds
Sands	Ferruginous earths	Dolomites	Coals	Cherts	Anhydrite beds
Breccias	Mudstones	Oolites		Diatomites	Potash beds
Conglomerates	Shales	Travertines			Magnesia beds
Sandstones	Sandstones				Nitrates
Quartzites					Phosphates
Arkoses					
Silts					
Flagstones					

←— May be formed as Fragmental (i.e., by mechanical processes), chemical or organic sediments —→
See also Chap. 2, Tables 2.1 and 2.2, Sections 2.2 and 2.4.

main groups (Table 3.7) according to their nature or major physical consti-tuents and then further subdivided into types according to their grain sizes and degree of compaction (or consolidation). Their mineral composition is so complex, being a combination of both primary and secondary minerals (the compounds resulting from both mechanical and chemical weathering, e.g., quartz and clay mineral particles) as to make a suitable classification into groups by mineral content again very difficult if not impossible.

The differences between sedimentary and igneous rocks is well illustrated by the difference between a conglomerate rock and a granite. A conglomerate (Latin: *Con* = together, *glomerare* = gather into heaps) is a collection of differently sized and mainly rounded particles formed from a variety of weathered rock fragments; these have been gathered into one place by natural processes and then bound together in a matrix of finer material, usually a cohesive clay, or an oxide or carbonate cement. A granite, as previously described, is formed however by the single operation of cooling a molten magma until crystallization occurs to produce a solid rock as one unit of interlocking mineral particles with quartz as a crystalline matrix.

(a) Arenaceous and rudaceous types

Sandstones are the most typical representative of arenaceous and rudaceous types and are solid rocks formed by the cementing, through natural agencies, of the grains of a sand. Thus the basic material of a sandstone is the same as that of sand, i.e., up to 95 per cent quartz (the most stable mineral), together with grains of the ferro-magnesian and silicate minerals, such as biotite mica and various iron compounds, as well as felspars, and often quantities of fragmented shelly rubble, all cemented together with either silica, iron oxide, lime ($CaCO_3$), or clay.

When mica is present in sufficient quantity, with its flakes arranged parallel to the bedding of the sand grains, the rock so formed is called a *micaceous sandstone*; similarly, sandstones derived from an igneous mother rock in their close proximity, may contain sufficient felspars to be termed *felspathic sandstone* or *arkose*.

Greywacke is a dark, greyish-coloured arenaceous sandstone, containing not only quartz grains but also plagioclase felspars and many other minerals, as well as angular fragments of varying sizes derived from other rocks like slate. These are usually old, coarse-grained rocks formed in regions adjacent to areas of igneous (principally granitic) rocks, as in southern Scotland and north Wales.

It will be remembered that felspars are weathered chemically into the clay minerals if transported in water for any distance from their igneous source rock, and these products form a major part of the resultant argillaceous sediments; thus felspar is not so common in sandstones except under the special circumstances mentioned previously in relation to *arkose*. Many sandstones (and sands) were formed by deposition of transported sediment in the shallow water of rivers, seas or lakes, either as marine, estuarine or lacustrine rocks, while others are simply the consolidated fragmental deposits of cemented wind-blown sand, such as those of the Permian–Triassic Periods (*see* Table 1.3). Some sands (and gravels) are of glacial origin from the Pleistocene period and others are recent beach and desert deposits.

The grain sizes in sandstone (and sands) may vary between 2 mm for the coarser-grained types to as little as 0·06 mm or less; when, however, this fine-grained lower limit of particle size begins to predominate, it is more common to designate the type of rock formed as a *siltstone*. (*See* Table 2.2.) As described in Chap. 2, the grain size of water-borne sediments is differentially sorted in the deposition process and this accounts for the granular uniformity of many sandstones, as well as the smoothness and rounded shape of their individual particles. (*Figure 3.7c.*)

The cementing material (as in a manufactured concrete) largely determines the resulting strength and weathering resistance of a particular stone. Thus a sandstone, with a clayey or calcareous cement filling the interstices between grains (calcareous sandstone), fractures or weathers readily through the cement, while the quartz grains are left projecting from the fractured surface and are themselves seldom cracked. Sandstones containing a silica cement are extremely tough and resistant to weathering; in these rocks, like quartzite, a fractured surface cannot be scratched with a knife, whereas a mark may often be made in the cementing material of the weaker types. When such a rock (e.g., *micaceous sandstone*) splits along its laminations into thin slabs of even thickness, it is termed *fissile*; these types were formerly a reliable source of the flagstones (or paving stones) of many English midland and northern towns, being much quarried from local carboniferous strata.

Rocks such as conglomerate, breccia, arkose, grit, etc., with particles of coarser grain sizes than those of the normal sandstone range are often listed under the heading *rudaceous*; while rocks containing particles at and below the lower end (0·01–0·005 mm) of the grain-size classification scale are usually extremely fine-grained and more argillaceous. Such a stone, which is often rather a sandy shale or siltstone than a sandstone and is formed from silt and clay as well as the finer sand particles, has a striped appearance and if its individual particles are hard enough, is crushed for use as a washed fine concrete aggregate.

Sandstones (and sands) are coloured according to the presence of iron oxide as cementing material, or as a thin film covering each grain; in ferruginous types, red and brown iron oxide is extremely common, especially on the outer surface of a rock, while in the mainly unoxidized interior, greenish or grey-iron compounds lend their colour to the rock, the grains of which may, however, be tinged with a red or brown colouring around their edges, through slight oxidation. *Calcareous sandstone* is the name given to any arenaceous rock in which the sand grains are surrounded by and cemented together with calcite ($CaCO_3$); the presence of this carbonate cement leads to the typical grey colour of these sandstones.

94

3.7. SEDIMENTARY ROCKS

Conglomerate consists of rounded pebbles or shingle cemented together and derived from water-borne igneous, sedimentary and metamorphic rock particles, the results of mechanical weathering. As described already, conglomerates are formed as shallow water or littoral deposits (*see Figure 2.20*) and where powerful currents have acted, may contain very large fragments; the matrix is generally of coarse, sandy material, but can vary from an extremely hard siliceous gel to a softish clay.

Various types are named according to the predominant kind of pebble found in them, e.g., *quartz conglomerate*, or *flint conglomerate*; *limestone* and *granite conglomerate*, etc. There are many deposits of red *Permian conglomerates* in Cumberland, the Midlands and S.W. England, and their material may be used for large engineering constructions as aggregate or 'Plums' (an engineer's colloquialism), in mass concrete retaining walls and sea defence walls.

Breccia is very similar to conglomerate (*Figure 3.7a*), but its constituent fragments are usually derived from untransported *scree*; these rock types are thus formed from angular particles cemented together, and consist of the coarse shingle material resulting from sub-aerial weathering of their source rocks, being mainly a type of consolidated *in situ* deposit. *Brockram* of Permian Age, found in the Eden Valley near Carlisle, is a typical example. There are similar igneous rock derivatives of pyroclastic origin known respectively as 'Tuffs' and 'Agglomerates'.

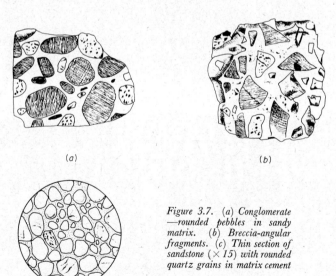

(a)

(b)

Figure 3.7. (a) *Conglomerate —rounded pebbles in sandy matrix.* (b) *Breccia-angular fragments.* (c) *Thin section of sandstone* (× 15) *with rounded quartz grains in matrix cement*

((c) Reduced one half on reproduction)

(c)

95

(b) Argillaceous types

Clay is perhaps the most typical argillaceous rock and consists of particles generally < 0·002 mm size, representing the finest land-derived sediments; these are produced mainly through the chemical weathering and transportation by water of the micaceous and kaolin type minerals so formed from Jurassic, Cretaceous and Lower Tertiary rocks (*see* Table 1.3) as well as from igneous rocks, the felspar minerals present being most susceptible to breakdown and decomposition.

Often, tiny comminuted quartz grains of fine sand or silt sizes (0·06 mm and less) are present in a sandy or silty clay. Other constituents may include calcareous, carbonaceous or organic matter and iron compounds, such as the sulphides and oxides, each of which has a predominant effect on the colouring of a particular variety. Clays are listed in Table 5.1 with references to their important properties of 'cohesion' and 'plasticity'.

Argillaceous sandstones are formed by the cementing of sand particles with a cohesive clay cement, although some calcareous types might be grouped here as they have a comparably low crushing strength. In some Midland varieties, calcium sulphate forms the cement and in certain other sandstones, mainly those found in Cheshire, barium sulphate is the major cementing material. All these types disintegrate easily through the weakness of their cements and are unsuitable for use as building stones.

Shale is a particular consolidated form of clay rock as noted in Chap. 2, which has 'fissile' properties and readily splits into thin layers, some of which may be slightly 'silty'. In 'paper shale', the bedding is so regular that the layers, each one of a slightly different texture, may be lifted and turned like newspaper pages.

Bituminous shale is rich in decomposed carbonaceous-organic matter and may be used as an inferior type of open-hearth fuel. Certain of these types are normally a source rock of 'oil'.

Mudstone is a non-fissile blue-black or grey coloured form of clay material, more compacted than 'shale' and generally without lamination or much evidence of plasticity; these stones break with an irregular conchoidal fracture. Some mudstones are formed in massive beds a foot or more thick, but others have a jointed and partly shaly structure, being interbedded often with clay shales; the mudstone termed 'bind' in coal mining areas, is commonly found in the shaly parts of the Coal Measures and especially so in the Midlands and northern England.

Marl, a calcareous clay or earthy mixture with at least 15 per cent calcium carbonate ($CaCO_3$), could be grouped with the calcareous rocks as a type of clayey limestone, although it is in most respects an argillaceous deposit; this well illustrates the difficulty of classifying all sedimentary rocks into distinct groups with very definite properties, since by increase of carbonate content, a consolidated marl becomes virtually a limestone.

Marling (or liming) is often adopted as a remedial measure for sandy and other agricultural top soils, in order to improve their vegetable-organic properties, consistency and degree of water-retention.

(c) Calcareous Types

These rocks consist mainly of calcium carbonate ($CaCO_3$), but some may contain up to 30 or 40 per cent of clayey material as, for example, when a marl blends into an argillaceous limestone, during their formation as sediments on a sea-bed. Limestones are more generally formed from masses of fossil shells, skeletal and organic remains; these accumulated, either in shallow sea-water—as occurred with the Mendips and Derbyshire strata— or in large freshwater lakes, which occurred, for example, during the formation of the Purbeck strata (*see* Table 1.3). Some of the fragmented shells and organic remains often dissolved in the water to form carbonates, especially when it contained carbon dioxide (CO_2) added from the atmosphere and had thus become a weak acid solution; later, through precipitation of this carbonate in the interstices of the compacted sediments, the natural calcite cement which binds the other material together, was formed.

Most limestones contain fossils by means of which their age may be readily determined and different varieties are often named after the particular organic remains which predominate in them—as, for example, in the case of 'shelly limestone' and 'coral limestone'. Many types contain more than 95 per cent of water soluble matter and are attacked by the carbonic acid resulting from the addition of atmospheric carbon dioxide to rain-water; the cold dilute hydrochloric acid (HCl) test, by causing effervescence of the rock, will always reveal the presence of calcium carbonate ($CaCO_3$), and rock fragments should be dissolved in hydrochloric acid to determine the amount of insoluble arenaceous/argillaceous or siliceous matter present in them, which may sometimes be considerable (over 30 per cent as noted above).

Limestones can easily be scratched with a knife to distinguish the presence of calcite from that of quartz (*see* Section 3.3—Mineral Identification); some types exhibit a smooth fracture and uniform structural appearance (as in a chalky rock), while others show a broken shell or rough surface structure (as in shelly or coral limestones, which consist largely of such remains set in a weaker carbonate cement). In all types, porosity depends on the degree of compaction and cementation of a particular rock and some have a very high porosity indeed.

Many limestones contain a small proportion of magnesium carbonate ($MgCO_3$), which is insoluble in cold dilute hydrochloric acid (HCl), unlike the calcium variety; mineral magnesium carbonate ($MgCO_3$) (Table 3.5), also forms the main constituent of dolomitic limestone, a rather special calcareous rock, often occurring with a highly crystalline dolomite content. The cold hydrochloric acid (HCl) test has no effect on such rocks but warm hydrochloric acid has, and this distinguishes them from other limestones. Magnesian limestones contain up to 20 per cent of magnesium carbonate, and are often termed dolomitic. It is prominent in the strata of Permian Age in Durham and extends southwards through Yorkshire to Nottinghamshire.

The coloration of limestones is dependent upon the amount of non-calcareous material or iron compounds present in them; and the appearance of individual rocks may vary in colour from white, through grey to yellow, red, or even black if siderite (iron carbonate) and other iron oxides are present in sufficient quantity.

Calcareous limestone is one of the organically formed deposits in which the cementing material is mainly calcite (calcium carbonate, $CaCO_3$), derived from the solution of broken shells present in most shore sands. The rock is usually grey in colour, and if the calcite content is excessive, the surface effervesces when cold dilute hydrochloric acid is placed on it; calcite cement fractures easily, and the crushing strength of such a rock is about 600 tonf/ft², as is that of a calcareous sandstone.

This stone like its sandstone equivalent is not suitable as external facing to buildings, since any acidulated atmosphere may dissolve its cement and cause disintegration of the outer surface. In particular, sulphuric acid present in town atmospheres reacts with the calcite cement to form crystals of calcium sulphate below the external surface, and any exposed face tends to scale; the same effect would apply to a calcareous sandstone, which might well be considered as a coarse type of limestone.

Such ordinary limestone rock has a fairly widespread distribution in Great Britain, especially in the Coal Measures and younger strata up to the Jurassic Period.

Oolitic limestones consist of a multitude of small spheres known as "ooliths', which are cemented together; each sphere about $\frac{1}{8}$ inch in diameter or less in size, is built up from a series of concentric skins of carbonate.

Oolite rocks sometimes contain iron compounds and the remains of shells; they are normally soft and unsuitable for use as a building stone in industrial atmospheres, although they are frequently used in southern and eastern England because of their availability and quality as a *freestone*.*

Oolitic types include the Bath and Portland stones found among the Jurassic to Cretaceous strata of the Mesozoic Era, although some oolite rocks may be deposited and interbedded with those of the carboniferous limestones from the Upper Paleozoic Era.

---

* 'Freestone' is quarryman's jargon for a rock which occurs in thick beds (Chap. 2) and can be easily quarried by cutting in any direction; this term applies particularly to the Bath and Portland limestones and also to certain sandstones found in Scotland.

Chalk—Is a very fine-grained, massive-bedded 'limestone' of rough texture, generally of great purity, but in which impurities may range from as low as 1 per cent to as much as 40 per cent of the total chalk content, as for example, in a white clayey or shaly marl; these impurities consist mainly of silt and sand or other forms of silica like 'flint' and 'chert', *see* Table 3.5.

Chalk rock is mostly a soft white carbonate which crumbles easily in water and washes clean after heavy rain, but some chalks are hard and fracture cleanly; it is suitable for making cement and is an important reservoir rock for water supplies (Chap. 6).

(e) Siliceous Types

Flint and chert are really forms of minutely cryto-crystalline* silica, which have formed into nodules; these are found mainly in the chalk strata, although chert often occurs in other limestone rock types such as Kentish Rag and the Hythe beds near Sevenoaks. Such nodules were probably formed from the remains of siliceous sponges, which dissolved in water and were subsequently re-deposited from solution as a 'gel'; very occasionally, evidence of fossil sponges and echinoids (sea urchins) may be found embedded in such nodules.

Concretions—Some flints may be regarded as an example of concretions, which are roughly spherical masses of material, foreign to a particular rock and varying from mere inches to a few feet in size; they are embedded in their surrounding rock, through which their substances have been carried in solution by percolating water, until collected into concretionary masses. For example, some iron and calcium carbonates found in clays or shales are often concretions, as are many *clay ironstones* from the Coal Measures.

Diatomite and diatomaceous earths—Diatomite resembles a consolidated white clay and is composed of the siliceous remains of microscopic algae named 'Diatoms', which accumulated in freshwater lakes; numerous sources of this organic sediment are present in Scotland and N.W. England.

Diatomaceous earth is composed predominantly of the skeletons from proterozoic sponges and sometimes sea organisms like 'Radiolaria' (Chap. 2) or simple microscopic plants; the organic deposit of such simple animal remains known as *Barbados Earth* is used as a fine abrasive and as an insulating material. Both diatomite and the less consolidated diatomaceous earth are inert and absorbent and are thus useful in chemical processes or as filter media, also for high temperature insulation, as a filler material in rubbers and paints, and as a cleansing agent. Fullers Earth and Bentonite are other special clays with important civil engineering uses, the latter in the construction of ICOS walls, for example.

* *See* Table 3.6—Textural terms.

(*d*) and (*f*) Carbonaceous—(organic) Table 2.1 and precipitaceous (chemical) Table 5.1, types of rock are mentioned adequately on page 57.

3.8. METAMORPHIC ROCKS*

These are probably the most complex of all rock types and are produced, as described in Chap. 2 by changes in the solid state of pre-existing igneous and sedimentary rocks; changes which are either physical, chemical, or a combination of both these effects.

There are varying degrees of *metamorphism*, these range from an effect that merely hardens sediments into rocks, which then remain easily recognizable from their original formation and mineral content (e.g., as in hardened sands or sandstones), to so complete a recrystallization that it becomes almost impossible to determine a rock's original state; in a glassy quartzite, for example, the mineral alteration has been so intense that few individual sand grains are distinguishable from their secondary quartz cement. Similarly, clay rocks may become *hard shales* of slaty appearance, but which continue to show their original fissile laminations; or they may be so altered by a pressure metamorphism as to become true slates with crumpled or obliterated bedding planes, and possessing the slaty-cleavage mentioned in the following paragraph.

More intense metamorphism produces the typical *schistosity* of mica and hornblende schists, by formation of their secondary minerals into the soft fissile layers along which these rocks may be easily split; the wavy appearance of their foliation planes, however, enables one to distinguish schists clearly, when they are compared to the straight laminations of a sedimentary, or even a partly metamorphosed, shale-slate rock. The final stage of metamorphism is shown in the *Gneisses*, in which none of the original rock structure remains physically intact or distinguishable, whether it was formally that of a sedimentary or igneous type; from the latter family alone, *serpentine* for example, is a typically complete metamorphic derivative —that of a totally altered olivine *peridotite* rock.

Most metamorphic rocks are hard and, if of sedimentary origin, usually devoid of their original fossils, which have been obliterated by the alteration process; such rocks are very apparent in the Highlands of Scotland where they cover and are exposed over wide areas of land surface. Six important examples are now to be described.

Gneiss is generally formed by dynamic metamorphism of coarse-grained igneous rock layers and is similar in composition to a granite; the metamorphism leads to a 'branded' appearance as the light and dark minerals are arranged in folia, alternate layers of quartz and felspar with mica between them.

* From Greek *Meta* = change of. *Morphe* = shape. These types are not very common or widespread in England.

Schist is a similar more finely foliated rock formed either from basic lavas or argillaceous sediments and examples contain mainly the mineral compound designating the name of a specific type, e.g., mica schist and hornblende schist. A schist will split easily along its wavy foliation planes, and pieces rather like the irregular layers of puff pastry break off to expose a glistening flaky mica surface between them; many of the quartz grains present have been elongated by the pressure metamorphism of schist formation and thus they lie in the planes of schistosity.

Gneisses and schists are often contorted, faulted and disturbed, and to the uninitiated, their outcrops may give the appearance of a banded sedimentary rock. Both occur over wide areas of the Scottish Highlands, Anglesey and Cornwall. When of sedimentary origin, they have a lower felspar content and are then called para-gneisses or para-schists to distinguish them from the ortho-gneisses and schists of igneous origin.

Quartzite originates from sandy sediments and is often formed from arenaceous/argillaceous rock metamorphosed in the aureole zone of basic intruded lavas; when cemented by silica, these baked sandstones may be transformed into very tough, hard quartzites such as glassy quartzite or even into para-gneisses. Not all hard and tough quartzites, however, are of metamorphic origin, as there are also sedimentary rock types, like certain sandstones with a strong enough silica cement, which are considered to be almost in the same category.

Slate—Clay sediments are formed into *mudstones* and *shales* by normal sedimentary compaction; thence into *slates* by the intense pressure of dynamic metamorphism and sometimes even further transformed into *micaschists* or *gneisses*. A slate possesses sets of cleavage planes (its slaty cleavage), which are distinct from the original sedimentary laminations and thus slates usually break into thin sharp splinters or flat sheets/slabs of rock, when hit carefully with a hammer and these pieces can be used as roofing or facing material for buildings.

The Palaeozoic sediments of north Wales contain many famous slate quarries like those at Penrhyn, Llanberis and Nantlle, especially in the Cambrian and Ordovician rock series (Table 1.3).

Marble originates from a pure limestone which has had its calcite content recrystallized and any fossils destroyed as a result of a metamorphosing heat and/or pressure, usually emanating from an intruded igneous rock magma. Marbles are thus composed of highly crystalline calcite, often pure white in colour like the Carrara marble of Italy and with an even-grained textural appearance. Limestone may also be metamorphosed into a 'calcareous gneiss', but many so-called building or commercially used marbles are actually polished limestone; for example, the Ashburton marble, a 'Devonian coral limestone' from Derbyshire, and Purbeck marble of the Upper Jurassic system, both of which contain fossils. Most decorative marbles, however, are produced in Italy and Greece, including its many islands.

Anthracite is a metamorphosed coal in a highly dense mineralized state, owing this special property to the effects of heat and/or pressure on the

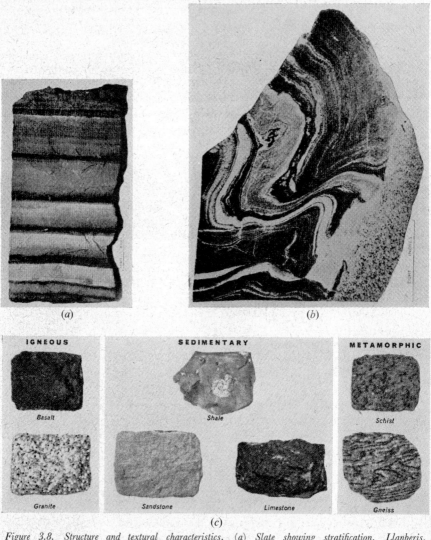

(a) (b)

(c)

Figure 3.8. Structure and textural characteristics. (a) Slate showing stratification. Llanberis, Caernarvonshire. (b) Contact of granite and folded gneiss, Tillyfour, Aberdeenshire. (c) Various

((a) and (b) Crown copyright Geological Survey Photograph. Reproduced by permission of the Controller, H.M. Stationery Office)

structure of ordinary bituminous (organic) coal. It possesses by far the highest carbon content (94 per cent) of any type of solid fuel and burns smokelessly, since volatile constituents form less than 10 per cent of its total weight.

3.9. CONSTRUCTIONAL MATERIALS

Building Stones: General Remarks on Different Types; Engineering Properties, Usage

Certain varieties of rock are eminently suitable for building purposes although almost any type of igneous, sedimentary and metamorphic rock, if reasonably compact and cemented, may be used as a local source of building stone, if readily available.

The really important properties required for any building stone are those of compressive strength and durability to weathering, while the quarryman considers bedding, jointing and cleavage, along with uniformity and soundness for 'dressing', to be prime considerations. The compressive strength of most rocks is seldom inadequate for average building needs, but durability to the corrosive weathering and chemical action of the atmosphere in cities and industrial towns may be lacking, although some chemical weathering of the outer rock surface is necessary to form a protective weathered film on the stone. Since the mineral particles of a stone are usually hard enough, durability is largely dependent on the degree of compactness, close knitting of particles and the strength of any cementing material. For example, loosely cemented sandstones are especially friable, although quartz is the predominant mineral constituent, whereas a compact limestone consisting mainly of mineral calcite with a calcite cement may be suitably hard and durable to weathering.

A uniform medium-grained rock is the most suitable building stone, because it dresses and weathers evenly and has the best resistance to corrosive weathering, as well as frost, temperature change and insolation. The porosity is also of supreme importance, for through open pores rainwater is absorbed into the stone and its presence may lead to disintegration of the surface by frost action or acid attack. In finely porous rocks, water is held in the pores by capillary attraction and such a stone is liable to remain saturated, being thus more susceptible to frost action. In the same way, its original *connate water* of high mineral content is often present in any deep-quarried or mined porous rock; when quarried, the 'green stone' is dressed and carved before this so-called *quarry water* can dry out at the surface to form a seasoned crust on the weathered face of a naturally matured stone.

The jointing in igneous and metamorphic rock types, or 'jointing' and 'bedding' in sedimentary rocks, largely determine their value as building stones; e.g., whether a particular rock is suitable as a *freestone*, or unsatisfactory, except for use as road stone, concrete aggregate, etc. Even so, care must be taken to lay a freestone with its bedding planes horizontal when constructing a building, otherwise these planes of weakness may be in the direct line of compressive stresses induced after loading and the rock made liable to rapid opening or splitting, even when subjected to only normal weathering.

Thus, practically every rock which may be quarried or dressed, has a crushing (or compressive) strength more than adequate for normal building work; although transverse (bending or shear) strength and tensile strength, where stones are to be used as 'lintels', are not necessarily so satisfactory and generally bear no relation to their crushing strength. Limestones are the most used rock for building purposes in Great Britain.

The former D.S.I.R., through the Building Research Station and Road Research Laboratory, established standards and methods of test for building, engineering and road materials; various B.S. specifications and Codes of Practice describe the standard methods of testing rocks or stones for strength, porosity, durability, etc., and the microscopic examination of their sections for mineral composition and texture. The Codes in particular give any relevant data required, and advise on the best use of material in engineering design and construction; *see* for example, B.S. 435, 706, 802, 812, 882, 1198, 1278, 2847, 1377* and *Site Investigations* C.P. 2001, etc.

Since limestones are the most used rocks for building purposes in Great Britain, a special catalogue and classification by geological strata of their sources, properties and uses on constructional work should be attempted by the interested student.

Road stones—The properties looked for in a suitable building stone are also necessary in any good road metal but with special attention being given to compactness, resistance to traffic abrasion† and natural weathering. Dolerite is perhaps the most important example of a road stone because its pieces readily hold a coating of tar and bind together firmly, and it also possesses the properties mentioned earlier. The 'Great Whin Sill' is a most prolific source of dolerite road material, quite apart from its long-established history as a quarried or mined building stone since the Romans used this rock for constructing parts of 'Hadrian's Wall' in Northumberland.

However, any similar rock which possesses close and irregular joints can be useful, since road stone must be crushed into particles which are then graded according to size into the typical 'chippings' employed as a top surface dressing; for this reason alone, fine-grained igneous types and hard limestones and quartzites, although from quarried rock with irregular jointing and thus unsuitable for building purposes, are often suitable road metals. Fine-grained basic igneous rocks wherever they occur in large masses are in great demand for macadamized road construction utilizing bitumen and tar binders. Thus, andesites, basalts, diorites and gabbros also, when

---

* B.S. 812: *Sampling and testing of mineral aggregates, sands and fillers.*
 B.S. 1377: *Soil testing.*
 B.S. 2847: *Glossary of terms for stone used in building.*
 † *See also* B.R.S. Digests Nos. 20, 21, 128 (1950) and Special Report No. 18 (1932).

available in quantity, are some of the best rock types for wear resistance under heavy traffic conditions. Although acid igneous rocks do not dress so well with tar as basic types, their chippings produce the optimum non-skid surface, so that granite chippings are well known on roads everywhere in Great Britain.

Of the sedimentary rocks, only certain Lower Carboniferous limestones produce reasonably strong material for compacted base hardcore and the denser Cambrian or Ordovician quartzites are often very good materials for a top surface layer. The main use of sedimentary rock types, however, is in providing gravel sand aggregate for the construction of concrete road or airway runway slabs and pavements. In this connection the following is worthy of note: a new map at 10 miles to 1 inch scale (1/625,000), published by the Ordnance Survey (Sept. 1965) as part of the national planning series for Great Britain, shows the locations of a total of 850 gravel-sand workings and borrow pits (graded according to three sizes) in England and Wales, for the benefit of producers, civil engineering and building organizations and others. Gravel and sands together are produced today at a rate approaching 100 million cubic yards annually, and are thus second only to coal as an economic product in both output and value; suitable sands are found in the Thanet sand strata of the London basin and the Bagshot sands of Hampshire (both tertiary), the Greensand and Wealden of Kent, Hampshire and Wiltshire (cretaceous), the Bunter Sandstones of the midlands and N.W. England (triassic) apart from glacial and modern river/estuarine sands, such as those dredged increasingly from the sea in the outer Thames estuary.

Sands, clays, brickearths, cements, etc. economic uses in general—sands are required for making cements, mortars and concretes; for use in the foundry moulds into which molten iron is cast and for the sand-blasting of components: for the manufacture of silica bricks used as furnace linings and likewise for glass-making* and as a filter material in water purification beds, to quote but a few examples. Clays are necessary both for common brickmaking and for firebricks, tiles, drains, porcelains and refractories, but for many refractory furnace linings, quartzite and dolomite are used; limestone is heated to manufacture the quicklime used with sand in mixing mortars, but the mixture of one-third clay and two-thirds limestone which is burnt to make Portland Cement, forms perhaps the most important building and civil engineering product of these materials; likewise, Plaster of Paris is manufactured by heating gypsum, to boil off its water of crystallization.

Conclusion

The foregoing descriptions of various elements within the sphere of constructional and industrial materials of interest to the engineer are necessarily abbreviated, but many excellent books contain detailed accounts

* *See* B.S. 952.

for the student who is more concerned with these aspects of economic geology. Suitable references are listed (*see* page 109).

3.10. EXERCISES

Igneous rocks

1. Give in outline the classification of igneous rocks and discuss the criteria on which their mode of classification is based.

In particular, describe the mineral composition, textural variations, physical properties, economic uses and occurrences of three principal *Acid* igneous rocks of engineering importance.

2. Give an account of the life history of a porphyritic granite from the time of its intrusion to its final destruction during weathering under humid conditions. (From I.C.E. Pt. II examination Apr. '62.)

3. (*a*) Describe how you would distinguish a rhyolite from a basalt, and a granite rock from a gabbro. How do the andesite and diorite rocks compare in texture and composition with the above-mentioned types?

(*b*) Suggest various locations for these rock types in the British Isles and relate the particular modes of occurrence which have led to their formation as distinctive rocks.

Sedimentary rocks

4. (*a*) Describe the tests you would carry out to distinguish between a pure quartzite, limestone and dolomitic limestone, and list some sedimentary rock types in which the minerals quartz, calcite, felspar, augite and olivine commonly occur.

(*b*) Briefly outline the character and properties of the following rocks, giving examples of the regions in the British Isles where typical examples may be located:

(*i*) Shale and slate; (*ii*) Flint and chert; (*iii*) Oolitic limestone; (*iv*) Flagstone; (*v*) Argillaceous sandstone; (*vi*) gypsum.

5. (*a*) Describe concisely, the important sedimentary rocks used in civil engineering and building. Mention localities where adequate supplies of these rocks are available and indicate the use for which each type is especially suitable.

(*b*) Account particularly for the main rock materials used in building activities in a locality known intimately to you, and compare their relative advantages and disadvantages. (From I.C.E. Pt. II examination, Apr. 1951.)

6. *Either:*

Define the term 'arenaceous rock'.

Write an account of the mineralogy of the main types of arenaceous rocks and give examples of arenaceous formations in Great Britain or in any country with which you are familiar. (From I.C.E. Pt. II examination, Apr. 1960.)

Or:

Give a general account of sandstones, emphasizing their mode of formation, their variations in composition, sources of supply in Great Britain, and economic uses.

Metamorphic rocks

7. (*a*) Describe the phenomenon of metamorphism, and explain the difference between metamorphic and sedimentary rocks.

(*b*) Give as examples the metamorphic types likely to be derived from: (*i*) shale; (*ii*) granite; (*iii*) pure limestone; (*iv*) sandstone; (*v*) coal; and outline briefly the properties of these rocks as engineering/building materials, referring to their sources of supply in Great Britain. (From I.C.E. Pt. II examination, Oct. 1952.)

8. (*a*) Distinguish between the various degrees of dynamic and thermal metamorphism in producing rock types ranging from hardened sediments to completely recrystallized rocks, illustrating by suitable examples the variety of changes that may occur.

(*b*) In particular, describe the history of formation of a slate from its origin as a sediment to its final metamorphosed form.

General

9. (*a*) Write brief notes on the origin, texture, mineral content, physical properties and uses, of the following rocks: (*i*) basalt; (*ii*) marble; (*iii*) argillaceous sandstone; (*iv*) schist; (*v*) oolitic limestone; (*vi*) rhyolite; (*vii*) conglomerate.

(*b*) Describe how the angle of slope of laminations like those present in a mica schist would be liable to affect the excavation programme during construction of: (*i*) a reservoir; (*ii*) a tunnel.

10. (*a*) Discuss the requirements desirable in a good constructional stone, mentioning briefly the main rock types used for building purposes. (*b*) By comparison mention the particular properties relevant to a road stone.

Mineralogy

11. Describe briefly the physical tests to be applied in attempting to identify a group of unclassified minerals, illustrating each test by reference to appropriate mineral examples.

12. Write brief notes on the following lists of rock-forming minerals, including their chemical constituents, crystalline form and types of rock in which they occur:
either (*a*) quartz; (*b*) olivine; (*c*) orthoclase felspar; (*d*) gypsum; (*e*) calcite
or (*f*) muscovite; (*g*) kaolin; (*h*) hornblende; (*i*) haematite; (*j*) rock salt.

13. (*a*) Discuss the definition of a rock as an 'aggregate of minerals', with special reference to (*i*) granite; (*ii*) quartzite; (*iii*) coal; (*iv*) rhyolite; (*v*) limestone; (*vi*) clay.

Illustrate your answers with particular attention to the distribution of the mineral 'quartz' in rocks. (From I.C.E. Pt. II examination, Oct. 1952.)

(*b*) Refer briefly to the action of water in causing the disintegration of rock masses and the formation of 'secondary minerals'.

3.11. PRACTICAL WORK—REFERENCES

Examples are recommended for display to student groups, or for inspection by individual students as explained in Chap. 2, section 2.8, pp. 63, 65. Use the 25 miles to the inch geological map of the British Isles to locate areas of igneous, sedimentary and metamorphic rocks. The larger scale, 10 miles to the inch (1 : 625,000) map of Great Britain supplied in two half sheets could also be utilized.

Exhibits at the London Geological Museum of the I.G.S. Petrology, mineralogy and resources

Exhibits of precious, semi-precious and ornamental minerals are described in the *Guide to the Collection of Gemstones*, as well as such ordinary examples as quartz and other forms of silica. Ornamental stones like Cornish Serpentine and Derbyshire Blue John are also displayed. Information can be abstracted from the *Short Guide to the Exhibits*, Dioramas and picture postcards and also colour transparencies, are available for certain exhibits and also particular mineral and rock samples—*see* typical references *MN, C* and *CT* as follows.

For example:

British ornamental stones and marbles

	Postcard No.
Geological globe showing distribution of igneous, sedimentary and metamorphic rocks	
Earth Structure globe	(MNL532)
Penrhyn slate quarry	(MNL500)
Portland stone quarry	(MNL506)
Quarry in Permian sandstone, Dumfriesshire	(C3596)
Granite quarry, Kemnay, Aberdeenshire	(C3753)

Rocks under the microscope

Thin slices and transparent colour photographs showing the appearance of rock sections in polarized light.

Igneous, sedimentary and metamorphic rock and minerals exhibits under British Regional Geology headings.

British mineral collection.

British building stones including sandstones, limestones and slates.

Small scale maps are available showing: The coalfields of England and Wales. The iron ores of England and Wales. The limestones of England and Wales.

3.11. PRACTICAL WORK

Exhibits and postcards from the British Museum, Natural History Department, as listed in their NHM Form 170 and Form 22A, are available on the same basis as those of the Geological Museum. For example: Postcard Sets D1–D10* including Precious stones and gems Sets D3–D6.

REFERENCES AND BIBLIOGRAPHY

Bowen, N. L. (1963). *The Evolution of the Igneous Rocks*. London; Constable

Daly, R. A. (1933). *Igneous Rocks and the Depths of the Earth*. 2nd edn. New York; McGraw-Hill

Dana, E. S. and Ford, W. E. (1932). *A Textbook of Minerology*. 4th edn., New York; Wiley

Evans, J. W. and Davies, G. M. (1967 Reprint). *Elementary Crystallography*. 2nd edn. London; T. Murby

Guppy, Eileen M. and Sabine, P. A. (1931–54). *Chemical Analysis of Igneous Rocks, Metamorphic Rocks and Minerals*. H.M. Geological Survey and Museum Sectional List No. 45—General Memoirs

Hatch, F. H., Wells, A. K. and Wells, M. K. (1952). *The Petrology of the Igneous Rocks*. 12th edn. London: T. Murby

Hatch, F. H. and Rastall, R. H. (1957). *The Petrology of the Sedimentary Rocks*. 3rd edn. (revised M. Black). London; T. Murby

Jones, W. R. and Williams, D. (1954). *Minerals and Mineral Deposits: A Conspectus*. Oxford University Press

Knight, B. H. and Wright, R. G. (1948). *Builders Materials*. London; Arnold

Knight, B. H. and Wright, R. G. (1948). *Road Aggregates, their Uses and Testing*. London; Arnold

Moorhouse, W. N. (1959). *Study of Rocks in Thin Sections*. New York; Harper

North, F. J. (1929). *Limestones; their Origins, Distribution and Uses*. London; T. Murby

Phemister, J., Guppy, E. M., Markinch, A. M. D. and Shugold, F. A. (1946). 'Roadstone: Geological Aspects and Physical Tests'. *D.S.I.R. Road Research Special Report No. 2*. London; H.M.S.O.

Read, H. H. (1953). *Rutley's Mineralogy*. 25th edn. London; T. Murby

Read, H. H. (1957). *The Granite Controversy*. London; Allen and Unwin

Read, H. H., Sutton, J. and Watson, Janet. In Press. *Metamorphic Geology*. London; T. Murby

Schaffer, R. J. (1932). *B.R.S. Special Report* No. 18. 'Weathering of Natural Building Stones.'

Shand, S. J. (1947). *Useful Aspects of Geology*. London; T. Murby

Shand, S. J. (1951). *The Study of Rocks*. 3rd edn. London; T. Murby

Shand, S. J. (1959). *Eruptive Rocks*. 5th edn. London; T. Murby

Singleton Green, J. (1946). *Limestone Concrete*. London: Chapman and Hall, London; H.M.S.O.

Smith, H. G. (1956). *Minerals and the Microscope*. 4th edn. London; T. Murby

Spencer, L. J. (1932). *The World's Minerals*. 3rd edn. London; (T. Murby) Allen and Unwin

Valton, P. A. (1946). *Limestone Roads*. London; Chapman and Hall, London; H.M.S.O.

Zim, H. S. and Schaffer, P. R. (1965). *Rocks and Minerals*. London; Hamlyn

* Set D1 contains 10 cards showing specimens of meteoric stone and iron.
  Set D2 contains 5 cards showing specimens of meteorites, which have fallen in Great Britain.
  These are relevant to the discussion of Earth Structure (Chap. 1) and should be compared to typical crustal rocks.

CHAPTER FOUR

STRUCTURAL GEOLOGY

4.1. INTRODUCTION

In the preceding chapters, a description has been given of how sedimentary rocks were laid down in more or less horizontal beds to form laminated *rock strata* on top of the earth's generally harder and earlier formed crust. Some *strata* may subsequently have been disturbed to form *folds* or undulations, others having been tilted so that their beds now possess a *dip*.

Figure 4.1. Quarry in Permian sandstone: Lochabriggs, Dumfriesshire

(This is one of the principal Scottish building stones. The view shows the quarry face and working platform: rough blocks are being dressed at top of quarry.)

(Crown copyright Geological Survey photograph. Reproduced by permission of the Controller, H.M. Stationery Office)

Frequently, *joint planes* are present also in both sedimentary and igneous rocks (*Figure 4.1*) and have been mentioned as being of great value in quarrying such rock, when they largely determine the size of blocks actually quarried. In general, the more consolidated rocks have the most definite joints and it is thought that jointing is due partly to shrinkage and partly due to earth movements occurring during the deposition and sub-sequent compaction of a sedimentary material. As will be seen later (Chap. 6), the flow of underground water takes place mainly through joints and fissures in rocks, that otherwise would be completely impervious to water.

The side faces of joint planes often show signs of their rubbing together, due to a relative crustal movement, by the appearance of 'slickensides'; these are parallel grooves or scratches, which may be vertical, horizontal or inclined, occurring in whole series on the smoothed faces of rock joints, fractures and fissures and they indicate the direction of original movement. Some joints, called 'master joints', in clays and shales, cut through many strata and large thicknesses of rock, while small discontinuous jointing is more common in sandstones and limestones. In conglomerates, joint planes pass clean through even the hardest pebbly material and follow a more or less straight path.

We shall now discuss the whole spectrum of Structural Geology (Note: not individual rock structures and properties as in Chap. 3) as it relates to the natural forms of land surfaces and connect these to their underlying geological origins in the *stratification, folding, faulting*, etc., of sedimentary rocks together with the other major structural effects due to igneous and metamorphic rock types. Finally, the construction of geological maps and the basic principles involved therein are dealt with.

4.2. STRATIFICATION AND THE BEDDING PLANE

Dip and Strike

Beds of rock are bounded by their 'bedding surfaces', upper and lower. In the simplest cases of continuous deposition (Chap. 2), the bedding surfaces of a series of rock beds (or strata) are parallel and these beds then form a 'conformable series', i.e., they were laid down regularly one above the other, the lowest in order being the oldest, as indicated in Table 1.3. Each bedding surface is common to two beds, being the 'lower bedding surface' of the upper bed and the 'upper bedding surface' of the lower bed. In the simplest and ideal case, such a bedding surface can be considered as a 'bedding plane'. (*Figure 4.2.*)

The *dip* of a particular stratum is the direction and angle to the horizontal of the line of greatest slope of any 'bedding plane' in the stratum, i.e., the maximum slope angle possible for such a plane. It may be expressed in degrees or as a tangent (e.g., a gradient of, say, 1/10). This value is known as the *true dip* and is at right angles to the *strike*, which is the direction of any horizontal line on the bedding plane (a direction called by miners, 'the level course'). On a geological map the 'true dip' direction of any 'stratum' is indicated by an arrow with the amount of dip shown beside it and referring to the dip value of the point of the arrow. Horizontal bedding is indicated

111

thus, ⊣⊢, and vertical bedding thus, —— | ——, the longer line showing the 'strike' direction.

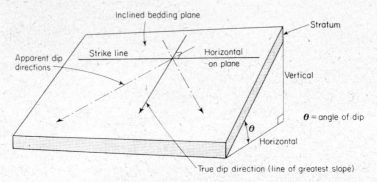

Figure 4.2. *The bedding plane*

A *strike line* which is also called a *stratum contour* is naturally a 'contour line' of the bedding plane. A series of such lines at a fixed vertical interval will be the contours of this plane surface and on a map will be projected on to the horizontal plane of the map. Thus straight, parallel, equidistant strike lines at the same vertical interval clearly represent a flat inclined plane, the 'dip' of which is easy to deduce from the ratio of vertical interval to horizontal distance between strike lines; both, however, must be in the same units as measured from the scale of the map, and the answer is then obtained as an 'angle of dip' or a 'gradient'. Similarly, the 'stratum contours' of a 'bowl-type' structure or a 'dome' would appear as a series of concentric circles.

The *apparent dip* in any given direction is the angle to the horizontal of a sloping line taken in that direction on the bedding plane; this angle is always less than that of the true dip and its limiting values may vary from zero (when it becomes a strike line) to that of 'true dip' (when it has a maximum value). True dip and strike are both fixed in value and direction

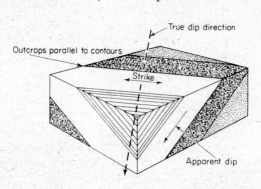

Figure 4.3. *Block model showing dipping strata with outcrop parallel to strike and apparent dip on sides*

for a bedding plane, but an 'apparent dip' value varies according to the particular direction or bearing taken for its line on the plane (*see Figure 4.2*).

Outcrops

When a stratum appears on the ground surface, it is said to *outcrop*, and this is shown by the line of intersection of its bedding planes with the ground; on a map, the upper and lower bedding planes are indicated by the horizontal projection of their actual outcrop lines to the scale of the plan view. A geological map is thus a plan showing the 'outcrop' of the several strata in the region represented, on which the areas occupied by the various strata are distinctly shaded or coloured (*Figure 4.3*). The width of any outcrop and the exact trace of their outcrop lines depends on two main factors: (*a*) the form of the stratum and (*b*) the ground topography (e.g., thickness and dip of stratum and whether the ground surface is horizontal or sloping, or a region of hill and valley undulations). Standard symbols are used to indicate the various geological phenomena encountered and are given in the Civil Engineering Codes of Practice, C.P. 2001, *Site Investigation*, Appendix C.

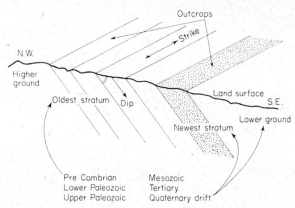

Figure 4.4. General arrangement of rock strata in Great Britain from N.W. to SE (dip direction)

Thus, if the 'outcrop' of the bedding plane is shown on a contoured map, the plane can be completely determined underground; even if only a small part of the whole outcrop is discovered in the field, the rest can usually be interpolated. Conversely, given the shape of the bedding plane underground, it requires only simple geometry to draw on a contoured map, the lines of 'outcrops' (*Figure 4.3*), although map interpretation generally involves the former problem.

If, say, the 25 miles to 1 inch geological map of Great Britain is studied, the outcrop of the rocks in eastern England from Dorset to Yorkshire shows their generalized 'strike' direction because the land is relatively level in relation to this distance and the strike coincides with the general trend of outcrops, although locally the outcrops and strike may not always agree in detail with this trend. In Great Britain generally, the rocks become progressively newer south-eastwards and their 'dip' direction may be inferred by the means demonstrated in *Figure 4.4* and confirmed by the generalized strike lines.

Because it is desirable for the students to become reasonably familiar with geological maps, the study of the main stratigraphical forms and their outcrops on 'block models', or of sketches in relation to the many outcrop variations caused by changes of ground topography, will first be necessary in order for him to develop a basic alphabet of three-dimensional mind pictures (i.e., the art of 'seeing solid') on which to build useful practical experience of map reading.

4.3. FOLDING

Sedimentary rock strata may have been disturbed by various earth movements. The simplest case of such disturbance is that of *folding*, where due to a slow lateral pressure in the earth's crust, the strata have formed undulations or simple *Folds*, which may vary greatly in size (*Figures 4. 5 a, b* and *c*).

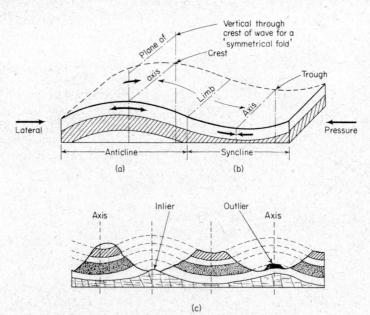

Figure 4.5. Folding. (a) Fold arched upwards as the crest of a wave. (b) Fold like the trough of a wave. (c) Partly denuded synclines separated by complete erosion of anticlines

114

Typical regional folds in England are the 'London basin syncline' and the Mendips and Pennines 'anticlines'. Folds with vertical axial planes, horizontal axes and similar dips on their two sides (or Limbs) are termed 'symmetrical', and those with unequal dips on their limbs are called 'asymmetrical'.

(a) *An anticlinal fold* (anti = against — cline) is one arched upwards, with its axis through the crest of the wave so formed; thus a ridge would be present initially on the ground surface, but weathering may have subsequently reduced it to a near level topographic feature, or the anticline may perhaps have been buried beneath newer rock formations (*Figure 4.5a*).

(b) *A synclinal fold* (syn = together — cline) is similarly a fold forming one of the troughs in rock undulations and thus would originally create a valley, which may well have been filled later by other sediments, so that the 'syncline' (or trough fold) itself has also been buried (*Figure 4.5b*). Anticlines are frequently worn down below the level of synclines, and often the only means of surface recognition is by the order of their stratum outcrops;* these show a near parallel repetition from the axes outwards—in the former case (a), with the older rocks at the centre near the axis and newer rock outcrops on the outsides and in the latter case (b), a precisely opposite arrangement occurs for synclinal outcrops.

(c) *A monoclinal fold* (mono = one) has basically a single undulation where the strata have dropped on one side only of the fold axis and consequently form a near vertical limb. Such a chalk fold runs through the centre of the Isle of Wight, dominating the island topography and ending in its south-west corner with the famous 'Sea Stacks' known as The Needles. Often a monocline develops into a 'fault' with the curled ends of the fractured strata denoting the direction of its 'throw'.

(d) *A dome* is a type of anticlinal fold with no axis and which dips circumferentially outwards all round, thus showing circular rock outcrops. A good example of this structure type is shown by the Harlech Dome of north Wales.

Overfolding

(e) *Overfolds* are caused in regions of intense lateral earth pressure. Their asymmetry is so large that often the axial plane is pitched right over and one limb may be overturned so that newer rocks become inverted under older rocks (*Figure 4.6*).

(f) *A recumbent fold* is rather like the welt of a shoe, where one fold has pitched over completely upon another, so that their axial planes lie virtually

---

* *See also* suggestions for sketching these later.

horizontally. Large areas of all these types of fold (*e*) to (*h*), occur in the Scottish Highlands and in any mountainous districts where peaks of alpine stature have at some past time been formed, although such movements probably occurred very slowly.

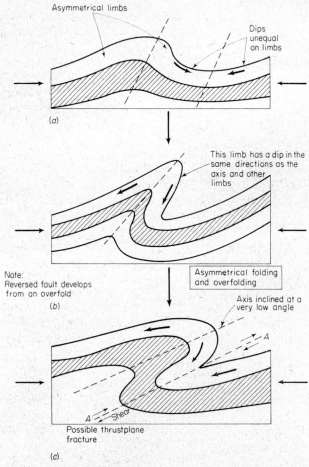

Figure 4.6. Over folds. (a) Simple folds of Figure 4.5 often develop anticlines and syclines with their fold axes tilted. (b) Further development of (a) into overfolds. (c) Overfolding under extreme thrust, may develop into a recumbent fold

(Note: An overthrust fault may develop from a recumbent fold by relative movement of strata along fracture plane A—A)

(*g*) *Isoclinal folds (and overfolds)* occur when the lateral earth pressure has been extremely intense and packs a series of folds together so that their alternate limbs are almost parallel and dipping steeply (*Figure 4.7*). Outcrops of worn-down beds forming 'symmetrical folds' when apparent on level ground are repetitive for each bed but equidistant laterally on either side of the fold axis and of equal bandwidth, whereas those of similar 'asymmetrical folds' and overfolds show a narrower width of outcrops on their more steeply dipping limbs.

Figure 4.7. Sharp isoclinal overfolding in siliceous schist (Moine series), at bridge over Blackwater, Little Garve, 1 mile north of Garve, Scotland

(Crown copyright, Geological Survey photograph. Reproduced by permission of the Controller, H.M. Stationery Office)

(*h*) *A pitching anticline* (*or syncline*) occurs when a fold is tilted so that its longitudinal axis lies at a dipping angle to the horizontal called the *pitch* (*Figure 4.8*).* This leads to the appearance of curved repetitive outcrops, which unite along the axis line on the ground surface, and if this is weathered down, they widen apart in a V-shape with increasing distance either side of the axis; in the case of a 'pitching anticline', the outcrops narrow together and the central older rocks disappear in the direction of the pitch, while in a 'pitching syncline' the newer beds disappear and the older outcrops widen in the direction of the pitch. (*See Figure 4.30d and e.*)

* See suggestions for sketching these later.

117

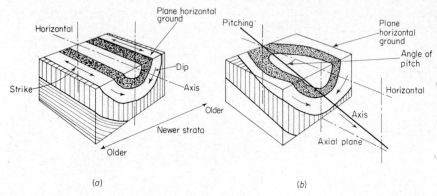

(a)   *(b)*

Figure 4.8. Folds. (a) Block model of denuded syncline showing outcrop repetition at level ground. (b) Block model of denuded pitching syncline showing outcrops

4.4. FAULTING

If a stratum cannot fold any further, it will *fault* under an extreme rapid lateral earth pressure or thrust, and a 'fault' is thus a fracture of the earth's crust along which relative movement of strata has occurred.

Figure 4.9. Normal fault. A normal fault in beds of sandstone and shale. The strata on the left of the picture have sunk along a plane fracture, and are brought into juxtaposition with dissimilar rocks on the right. Excavations at Mossend, Lanarkshire

Of the various fault types, the 'normal fault' is shown first (*Figure 4.9*) with a fault plane steeply inclined and the 'Hade' towards the downthrow side. Often masses of torn-off angular rock fragments from the adjacent fault faces fill any possible fault space or fissure opening in the ground and form 'Fault' (or Tectonic) 'Breccia', in conjunction with traces of 'slickensides'— polished, striated surfaces resulting from friction along the fault plane— to indicate the direction of relative movement, since such striations (or scratches) are parallel to the direction of fault movement.

The following terms are used in relation to 'faults' and are illustrated in *Figure 4.10*.

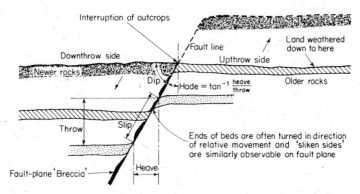

Figure 4.10. Normal fault. The fault plane has a steep, near vertical inclination and hades towards the downthrow side

(Note: Specify a 'fault' by Hade, side of downthrow and trend of fault plane outcrop —whether this is in the direction of strike or dip of the rock strata)

Hade is the angle made by the 'dip' of line or maximum slope of a fault plane to the vertical and is usually quite steep. The value of Hade is thus

$$\tan^{-1} = \frac{\text{Heave}}{\text{Throw}}$$

and may be expressed either as a tangent (gradient: 1 in 5 say) or as an angle in degrees.

Throw is the vertical distance between corresponding points of the same stratum on either side of a fault, and may vary in amount by any value from inches to several thousand feet. Faults with small 'throws' are often to be seen in the walls of any ordinary quarry.

Heave is the horizontal distance between two such corresponding points, and commonly measures less than the amount of any 'throw'; but contrari-wise, heave in a thrust fault usually represents the largest amount of relative movement between the corresponding strata. (See the following text.)

The presence of a fault is seen clearly on a map, generally as a coloured straight line outcrop (i.e., that of a nearly vertical plane intersecting the

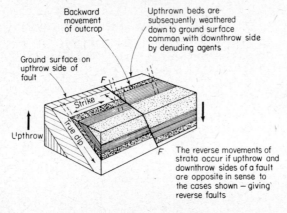

Backward
movement
of outcrop

Upthrown beds are
subsequently weathered
down to ground surface
common with downthrow side
by denuding agents

Ground surface on
upthrow side of
fault

Figure 4.11. Block model showing dipping strata, with outcrops parallel to strike—normal dip fault.

The reverse movements of
strata occur if upthrow and
downthrow sides of a fault
are opposite in sense to
the cases shown – giving
reverse faults

ground topography and running independently of the ground contours), with a small mark on the downthrow side (the side of the newer rock outcrops) and the interruption or disappearance of existing stratum outcrops.

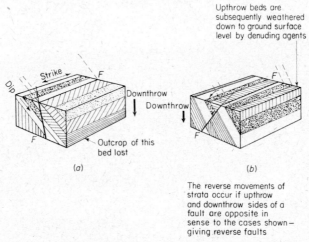

Upthrow beds are
subsequently weathered
down to ground surface
level by denuding agents

Downthrow

Downthrow

Outcrop of this
bed lost

(a)

(b)

The reverse movements of
strata occur if upthrow
and downthrow sides of a
fault are opposite in
sense to the cases shown –
giving reverse faults

Figure 4.12. Block models showing 'normal strike faults'. (a) Concealment of outcrops and beds. Throw in same direction as dip of strata. (b) Repetition of outcrops and beds. Throw against the direction of dip

A *dip fault* is one, the line of which extends in the direction of dip of the strata and which shows a characteristic interruption and repetition of outcrops somewhere else along the line of the fault (*Figure 4.11*).

A strike fault similarly extends linearly in the direction of strike of outcrops (*Figure 4.12*). Thus, the stratum movements may lead to concealment or repetition of outcrops according to whether the throw is in the same direction as the dip of the beds (as in a normal strike fault), or in the opposite direction to the dip of beds (reversed strike fault).

A *reverse fault* (or *overlap*) occurs when one set of strata have risen up over their corresponding strata, the Hade then being toward the upthrow side of a fault (*Figure 4.13*). Both 'normal' and 'reverse' faults occur as either dip or strike faults, so that these combinations can cause either 'normal dip' and 'normal strike' or 'reverse dip' and 'reverse strike faults', with consequent effects upon their stratum outcrops.

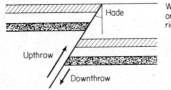

When Hade is at 45 degrees or more and one layer has ridden up over the other

Hade

Upthrow

Downthrow

Figure 4.13. Reversed fault (or overlap fault)

A thrust fault—In the case of near horizontal reversed faults with a large angle of Hade (called *thrust faults*), a stratum may override its corresponding stratum for many miles. Such faults are often a natural development from a fractured 'recumbent fold'. In north-west Scotland there are many examples of such 'overthrusts', the most famous being the Moine Thrust (*Figure 1.4*) in which Pre-Cambrian rocks have overridden Cambrian for a distance of 60 miles or more.

A *tear fault* may have no relative upwards or downwards movement of rocks, but shows evidence only of a relative horizontal movement of the same strata on either side of the fault (*Figure 4.14*). A *tear fault*, however, is more commonly combined with a normal strike fault, with the resultant effects on outcrops already described above.

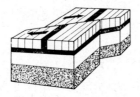

Figure 4.14. Tear faults

Step faults and trough faults—In many areas of faulting, *fault planes* tend to form groups in which the fault lines are more or less parallel. One such group of faults may form a series of 'steps' when their individual throws are in the same directions. Likewise, pairs of faults may lead to the formation

121

of a *horst* (or ridge) or to a *Graben* (or trough) when the simultaneous throws of each fault are similar, either of uplift or depression (*Figure 4.15a–d*).

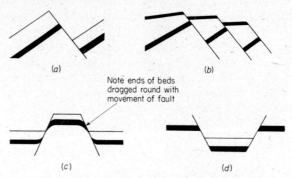

(a)

(b)

Note ends of beds
dragged round with
movement of fault

(c)

(d)

Figure 4.15. *Combined systems of faults.* (a) *Tilted block.* (b) *Step fault block.* (c) *Horst or ridge.* (d) *Trough or graben*

Referring again to the 25 miles to 1 inch geological map of the British Isles, the central valley of Scotland lies between two major fault scarps (the Highland Boundary Fault and the Southern Uplands Fault—see *Figure 1.4*) in the form of a 'trough' about 50 miles wide.

The famous Rift Valley of East Africa extending northwards from Tanzania through the Red Sea to the Dead Sea in Jordan is really a very narrow 'trough' between two nearly parallel fault plane 'horsts' and in the Dead Sea its floor has dropped well below the present day M.S.L. (Table 1.1). The Black Forest in Germany is similarly a typical 'horst' area, surrounding and rising above the 'Rift' valley of the Rhine in Westphalia. Both the Great Glen in north-west Scotland containing Loch Ness and the Vale of Neath, south Wales (Note: also Bala Lake in north Wales), are 'fault valleys'; i.e., the valleys rapidly corraded by erosion and denudation processes in the shattered and weakened rock material (mainly breccia) of their fault planes, especially when the valleys have been occupied by rivers which in course of time have captured the original regional topographic drainage system. (*See* Chap. 2—River Capture, etc.)

4.5. UNCONFORMITY

Unconformity occurs when a newer series of strata rest with discordance upon the denuded surface of older formations (*Figure 4.16*). The two series are bounded by a *Surface of Unconformity* which, whatever has happened since the unconformity was formed (*Figure 4.17*), remains as evidence of a time gap in the sequence of deposits. In most cases, the lower rock series were once dry land undergoing erosion when a subsidence allowed a re-inundation by water and the deposition of the newer series. The time interval involved may have been short or it may have lasted millions of years between the formation of two such independent rock series. Reference to their relative positions in the stratigraphical succession of the geological column (*see Table 1.3*) will elucidate this point more precisely.

Figure 4.16. Unconformity. Unconformable junction of Carboniferous Limestones upon Silurian flags. Comb's Quarry, Helwith Bridge, Yorkshire

(Crown copyright, Geological Survey photograph. Reproduced by permission of the Controller, H.M. Stationery Office)

The regional rock structure can be very complicated because not only may the lower series have suffered tilting, folding, faulting and denudation previous to subsidence, but both series may subsequently have undergone changes in their overall structure. There are many ways of recognizing an 'unconformity', i.e., difference in 'strike and dip' between the upper and lower series of beds; folding of the lower beds or similarly a faulting, both of which may stop at the upper series; the presence of pebbles, boulders and other fragments of denuded rock forming a 'conglomerate' at the boundary surface of the 'unconformity' and, on geological maps, often the clear evidence in different places of outcrops of newer beds resting upon, but

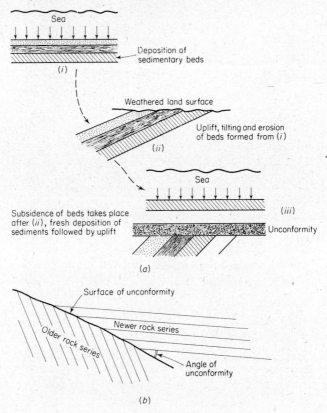

Figure 4.17. (a) Stages in the formation of unconformity. (b) An angular unconformity or overlap

widely at variance with, other underlying rock types. Most superficial deposits (i.e., 'Drift' like boulder clay, alluvium and peat) rest unconformably on their underlying solid rocks; hence the necessity for separate 'solid' and 'drift' geological map editions, as described later in Chap. 5, in order to avoid the complications which would follow in many areas of such

4.5. UNCONFORMITY

'unconformity', when related to the production and interpretation of a dual purpose drift/solid geological map.

Glacial moraines and other drift sediments have been adequately described in Chap. 2 and to this information the student is here referred for the present, although 'drift' rocks will be dealt with again in Chaps. 6–7 for their applications in mapping and engineering geology.

4.6. ESCARPMENTS, INLIERS AND OUTLIERS

'Escarpments' occur in a country where there are strata of successive beds, which generally include hard rocks like sandstone or limestone interbedded with layers of softer rocks such as shale or clay. The outcrops of the harder rocks stand out as ridges running in the direction of their strike, after weathering has denuded the softer rocks more rapidly to form a surrounding region with a lower ground surface, although the harder rocks must be interbedded between the softer ones if a typical *scarp and dip* topography is to develop (*Figure 4.18a*).

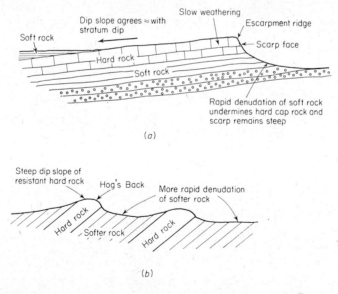

Figure 4.18. (a) *Typical escarpment topography.* (b) *Typical Hog's Back topography*

This type of scenery is prominent in the Chalk Downlands of south-east England and other hills of escarpment structure are to be found in many areas of Great Britain.

125

When alternate hard and soft beds dip steeply the differential erosion of their strata often leads to the formation of a Hog's Back ridge (*Figure 4.18b*), as has occurred on the line of outcrop between the towns of Guildford and Farnham in Surrey (south-east England).

Landslips caused by earthquakes* sometimes produce small random relative displacements of strata on either side of any 'fissure' so produced; for example, in Japan (1891) and Assam (1897) such movements occurred to form 'fault scarps' of 20 ft or more in height running straight across country irrespective of existing topography and rock strike, and in particular, that of 1909 at Madison, Montana, U.S.A., remains prominent, stretching for miles along the line of the earthquake fault. In England, the eastern edge of the Malvern Hills is an ancient eroded fault or scarp unbacked by a 'dip' slope (its distinguishing feature from ordinary escarpments) where the harder and older uplifted pre-Cambrian rocks abut against those of the faulted and softer new red sandstone.

Inliers and outliers—An *inlier* is an area of harder, older rock which is completely surrounded by newer rocks; an *outlier*, conversely, is an area of harder and newer rock surrounded by older outcrops.

Outliers occur when newer beds overlying older rocks have been denuded by a severe process of erosion (e.g., that due to river corrasion, or a local glacial action) in places down to the older rock; this process may generally be observed in slightly dipping or horizontal strata, and it leaves only isolated hills of the harder or newer rocks standing out above the regional topography.

Inliers, which may be formed from rock masses surrounded or criss-crossed by 'faults', are often found in valleys cut by rivers, or glaciers, again through slightly dipping strata or in denuded anticlinal areas (*see Figure 4.5*). Both these phenomena are important in providing an accessible study of rock types, which have been subjected to severe denudation, away from areas of their normal surface exposure. Outliers are often found in the region of an escarpment, when they form disconnected hill outcrops at some distance in front of the main 'scarp line' (or step feature).

4.7. IGNEOUS INTRUSIONS

Dykes and Sills etc.

Irregularities in the stratification of sedimentary rock may be caused by the intrusion of an igneous rock magma into and through fissures in existing overlying strata. (*See Figure 4.20.*)

A Dyke is a vertical or steeply inclined wall-like mass of solidified igneous rock with approximately parallel sides which has been injected through the overlying strata of sedimentary and other rocks. It may extend in width from mere inches to 100 ft or more, usually forming a sheet of considerable

---

* *See* Section 4.9 re earthquakes.

length and often cutting many miles directly across country, while the strata on either side are comparatively undisturbed, except for a local metaphorism of the surrounding country rock material.

At the ground surface most dykes either form a shallow upstanding eroded wall, when harder than their adjacent country rocks, or perhaps a trench when softer than the latter, having thus become the more rapidly weathered outcrop.

A Sill is a similar intrusion to a dyke, again of varying size in area and thickness, forming a horizontal or dipping sheet of solidified igneous rock between the bedding planes of its overlying strata, and likewise causing a localized metaphorism of any surrounding rocks. When passing from one stratum plane to another at a higher or lower level, it is termed 'transgressive' as is the Great Whin Sill of north-east England, which covers an enormous area, nearly all of it underground.

The rock composition in dykes and sills is mainly of fine-grained crystals forming typical micro-diorites, dolerites, trachytes, andesites and basalts which, due to their magma's cooling rates and contraction have often formed horizontal polygonal columns from side to side of dykes and similar vertical ones from top to bottom in sills. All intrusive igneous rocks are obviously younger than their surrounding rocks; there are great numbers (called *swarms*) of parallel running dykes and 'sills', many showing a typical columnar jointing *(Figure 4.19)* which traverse the country rocks of the western regions of southern Scotland, most of them having been intruded during the same geological period, as for example, tertiary volcanic rocks.

Figure 4.19. Columnar basalt—diorama. The island of Staffa—Eocene Period
(Crown copyright Geological Survey photograph. Reproduced by permission of the Controller H.M. Stationery Office)

Batholiths, Stocks, Bosses and Laccoliths

Batholiths form enormous masses of igneous rocks *(Figure 4.20a)* generally composed of granite or granodiorite types, with irregular dome-like roofs, and walls that plunge sharply downwards so that these vast intrusions enlarge with depth and appear to be without traceable foundations, at any rate by sub-surface boring. (*See* Chap. 1, page 14.) They occur most frequently in the heart of mountain systems and elsewhere in rock strata of all geological ages which have undergone intense folding, and are generally elongated parallel to the main trend of the mountain folding in which they are found. Often, and especially in low-lying moorland or semi-arable land, the rock surfaces of batholiths are ultimately exposed by denudation of their softer surrounding strata; they frequently become upland Tors, as on Dartmoor, Devon and Bodmin Moor, Cornwall, or, form as isolated mountains, e.g., in the Isle of Arran, south-west Scotland and at Mt. Leinster, near Dublin, Eire and also in north Wales and Westmorland. The name *Batholith* comes from the Greek (*Bathos*—depth; *Lithos*—rock or stone) thus meaning a 'depth rock'. There is still considerable debate among geologists as to the means of intrusion of 'batholiths', but their most probable origins are explained by such theories as those of 'cauldron subsidence' and 'magmatic stopping' (*see* Daly, 1938 and Read, 1957).

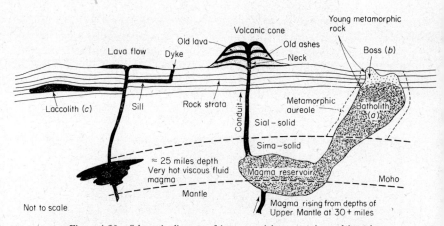

Figure 4.20. Schematic diagram of igneous activity—extrusive and intrusive

Stocks and bosses—Smaller separated intrusions of a similar type to the batholiths, but less elongated, and with their total plan dimensions covering only a few square miles in area are called *Stocks*. Many of these are probably offshoots of unexposed batholiths. When a stock has a roughly circular outline and steep sides, for example, like that forming the Shap Fell granite in Westmorland, it is referred to as a *boss* (*Figure 4.20b*).

Laccoliths—These are produced when an injected viscous magma is unable to spread between sedimentary rock beds into a large thin sheet to form a sill, but becomes instead a large enough mass to arch up the overlying strata as a mushroom-shaped dome covering the laccoliths' own top surface; a laccolith is often connected by a stem from its relatively flat base down to the magma reservoir where the laccolith material originated deep below ground (*Figure 4.20c*).

There are few good examples of real 'laccoliths' in Britain, though many stocks and bosses have been erroneously called laccoliths; but in the U.S.A. an outstanding example of laccolith formation is that of the Henry Mountains, situated in Utah, as part of the upper region of the Colorado Basin. It is worth noting that stocks are commonly discordant intrusions, whereas laccoliths are, like sills, generally concordant, with their overlying sedimentary strata. Laccoliths are found mainly intruded into horizontal strata and, although smaller than batholiths, they are still considered as 'major intrusions' with a rough diameter of several miles and formed mainly of granite type rocks.

The metamorphic aureole—The surrounding country rocks in contact with any igneous intrusion are commonly metamorphosed by the extreme heat of migrating fluids emanating from the magma, and 'metamorphism' of this kind has already been distinguished by name as 'contact metamorphism'. The zone of altered rock surrounding the intrusion, showing clear evidence of the great heat of the intruded magma, is called the *Metamorphic Aureole*. This metamorphism of adjacent rocks lessens progressively with distance away from the intrusion, until at some distance from it, the original country rock is completely unaltered. Around large 'batholith-boss' type emplacements, the pre-existing country rocks are usually metamorphosed for a mile or more, while those bordering on a dyke or sill intrusion are affected for only a short distance outwards from these phenomena.

Lava flows—In Great Britain there has been much extrusive volcanic action in past eras of millions of years ago, but no actual surface volcanoes, cones or craters remain evident—only the volcanic rocks, lava flows and their ancient " necks', described as follows. The volcanic hills of Antrim, N. Ireland, and the Inner Hebrides (e.g., Giant's Causeway, Staffa, etc., *Figure 4.19*) and also those of the Scottish Lowlands, the Lake District and north Wales, all stand out because of their relative hardness among their surrounding sediments. Ancient 'lava flows' are interbedded with sedimentary rocks, apparently as an integral part of them, whether horizontal, simply dipping, folded or faulted, and the lavas may be distinguished from intrusions like sills by the absence of any metamorphism in their surrounding strata, which were deposited after the solidification of these lava sheets.

4.8. VOLCANIC PLUGS (OR NECKS)

During eruption of a volcano, the gases accumulated under the internal pressure in the earth's outer crust are released and their outward rush clears the vent pipe of a volcano for the fluid magma to follow; ultimately, after sufficient eruptions, this vent invariably becomes choked with ash and debris or fills with a solidified 'lava flow'. The more silica rich and less fluid lavas, such as the finely crystalline rhyolite-dacite types, are often so highly viscous that they cannot flow away from the vent of the volcano and therefore eventually form a dome-shaped plug over the volcano's pipe. When this stiff lava protrudes down into the vent pipe and blocks it, the lava consolidates to form a *volcanic plug*, which often protrudes rather like a 'neck' after the main force of the original eruption has been expended.

Figure 4.21. Le Puy-en-Velay. Rocher St. Michael

These dome-type spines are typical of the Le Puy, Auvergne, regions of France (e.g., in the acid lavas of Puy-de-Dome and the Rocher St. Michel, a 'neck' of agglomerate, *Figure 4.21*, and they occur also in the U.S.A. at Mount Lassen, California. In the case of extinct volcanoes, such a plug is often exposed above the general surface level by weathering away of the softer surrounding strata; the 'plug' having a near vertical contact with its surrounding rocks and being independent in its outcrop of the regional topography as in the case of Arthur's Seat, Edinburgh, which forms a special rock feature of the landscape. Owing to the rapid cooling of the original lava which constitutes a 'plug', these have a finely crystalline acid composition—hence forming Rhyolitic-Andesitic rock types.

Volcanic plugs or necks are abundant in Central and Southern Scotland, standing out as round, sharp-edged, circular hills, as in Fife (Largo Law, etc.), in the Forth valley (Berwick Law, Arthur's Seat, etc.), and Morven, Argyll-shire—all of them the remains of ancient lava necks.

4.9. EARTHQUAKES*

Although in Great Britain earthquakes are rare occurrences, elsewhere they are often associated with volcanic activity and result from a sudden shock movement or break in the earth's crust after a long period of accumulating strain; the worst types of earthquake occur, however, independently of any volcanic eruptions. A severe earthquake may cause landslips and cracks to appear on the ground surface, and even produce an observable fault movement of many feet (e.g., a Fault Scarp); any sudden displacement of relatively shallow crustal material at a 'focus' depth of less than 20 miles causes an emanation of the multi-directional radial outward shock vibrations, described briefly in Chap. 1. The relative displacements of deeper crustal rocks, however (the seat of earthquakes), in actuality seldom reaches the ground surface and 'fault scarp' occurrences are rare events.

Generally, no features of any great geological importance result from earthquakes, but local damage to buildings and other engineering structures may often be severe. The worst damage is caused by 'L' type shock waves (see Table 1.2) which travel, as the slowest wave movement induced, through the sedimentary rock layers of the near-surface crust and have the maximum possible disruptive effect on buildings, services or similar works. Today, in earthquake zones and known shock regions, structures are usually designed to withstand all but the most severe of possible shocks; these zones affect mainly the circum-Pacific belt or 'Ring of fire'—see Chap. 2.

The 'Mercalli' etc. scales of shock and an outline of intensity measurements by seismograph of L, P and S waves, together with more details of earthquake phenomena, are given by Zumberge, 1931.

4.10. CONSTRUCTION OF GEOLOGICAL MAPS

The three-point problem

In general, a bedding plane may be completely determined if three non-linear points on it can be established both in position and in depth (or height) from a known base-line and reference datum. Clearly, any one strike-line and a point, or any two lines of 'apparent dip' will also furnish the same information.

For example, a bedding plane has a dip of 1 in 5 in a north-west direction and a dip of 1 in 4 in a north-east direction (*Figure 4.22*). These dips could

---

* Recent examples include the earthquakes at Skopje and Debar, Jugoslavia, in July 1963 and December 1967; Aggadir, Morocco in 1959, and others in Turkey (1967), Greece (1965) and Chile.
A paroxymal volcanic eruption of devastating effect occurred in September 1965 at Mount Taal, 35 miles south of Manila on the Philippine island of Luzon.

be measured with a clinometer if the beds are exposed above normal ground level either on a hillside or in the walls of a quarry. Two simple forms of surveying hand-clinometer are the Watney's—essentially a protractor type rotating circular pendulum—and the more refined Abney type, fitted with a level bubble, cross hairs, and a small vernier scale for vertical angle or gradient measurement.

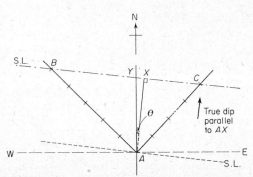

Figure 4.22. Solution to 3-point problem—apparent dip

Angle θ scales 6°20′ and AX scales 3·05 units horizontally

Solution—Draw on plain or squared paper, representing the horizontal ground surface, a line AB in a N.W. direction 5 units in length, to any suitable scale, and similarly draw AC in a N.E. direction 4 units in length, to the same scale.

Assuming A is a point on the bedding plane, then B and C are also points on the plane both at the same level of 1 unit below A.

Line BC is thus a strike line and AX may be drawn perpendicular to BC for the direction of True Dip; this has a bearing, which is measured from the tangent $\dfrac{XY}{AX}$ as N.6 degrees 20 minutes E. and an amount $I/AX = I/3·05$, where $\tan^{-1} I/3·05 = 18\frac{1}{3}$ degrees.

Thus the angle of 'dip' = 18 degrees 20 minutes at a bearing 6 degrees 20 minutes E. of N.

Example

Explain the difference between 'true' and 'apparent' dip of a stratum.

It is required to construct a building on a site whose average surface level is 85·00 O.D. (Ordnance datum). The surface soil is of poor quality but it is understood that a sound bed of ballast lies below it. Borings obtained from three neighbouring sites gave the following information:—

(1) 600 yards N. of site, ground level: 97·00 O.D.
Top of ballast 25 ft below surface.

132

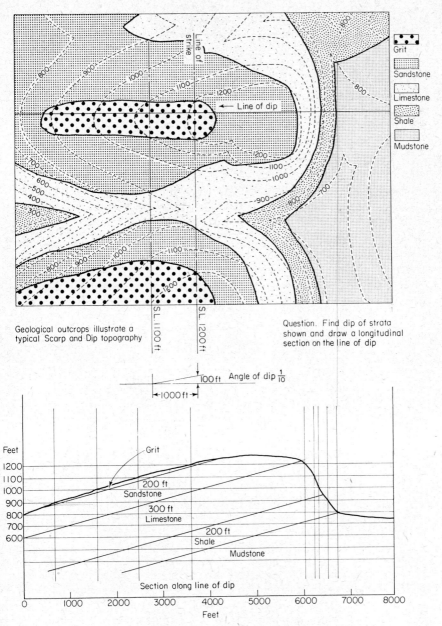

Figure 4.23. Dip and scarp structure

(By courtesy of T. Murby and Co.)

133

(2) 600 yards W. of site, ground level: 78·00 O.D.
>Top of ballast 32 ft below surface.
(3) 850 yards N.W. of site, ground level: 89·00 O.D.
>Top of ballast 27 ft below surface.

Assuming a uniform dip to the top of the ballast determine the dip and the anticipated level of the top of ballast at the site.

(*Ans.*: Dip = 1 in 95·4 at bearing S. 32°W. Level of ballast at site = 56·00 O.D.)

The Interpretation of Geological Maps

So long then as the outcrops (or borehole records) of three different points on it appear, any bedding plane can be completely determined. There are apparent variations of this, but they are still basically the 3-point problem. The procedure for solving a particular bedding plane problem is as follows:

Case 1—Note two points on the outcrop of the plane where it crosses the same ground contour line. A line through these two points is a 'strike line' of level value the same as that of the ground contour line. Note another point where the same 'outcrop' crosses any other suitable contour line; then draw a S.L. parallel to the previous one and of level value the same as that of this second contour. The bedding plane can thus be determined (as demonstrated on *Figure 4.23*) even if suitable contour lines need to be interpolated between the actual values of those given on the map.

If isolated outcrops are to be completed, S.L.s can now be drawn at the same vertical interval as the topographic contour lines to the boundaries of the map and the S.L. intersections with their corresponding contour lines marked. Finally, a smooth curve drawn through these intersections completes the outcrop over the whole extent of the plan area represented (as is demonstrated on *Figure 4.24*).

Case 2—More generally, three points may be found on the same outcrop where the topographic contour lines all differ in value. The bedding plane must be determined by a 3-point calculation before S.L.'s can be drawn and thereafter the problem is the same as in Case 1.

Note: If a stratum plane is horizontal, its outcrop follows the contour lines on the map exactly.

It should be remembered that the continuous outcrop of a bedding plane like ground contour lines need not close on itself within the limits of a map, as happens in *Figures 4.23* and *4.24* with most outcrops. It may also appear as detached pieces but after these are traced to completion, an outcrop must ultimately close on itself even if this final closure is beyond the edges of the mapped area. By measuring the appropriate horizontal distances between S.L.'s both the 'true dip' and the 'apparent dip' in any given direction such as *cb* on *Figure 4.24*, may be established.

4.10 CONSTRUCTION OF GEOLOGICAL MAPS

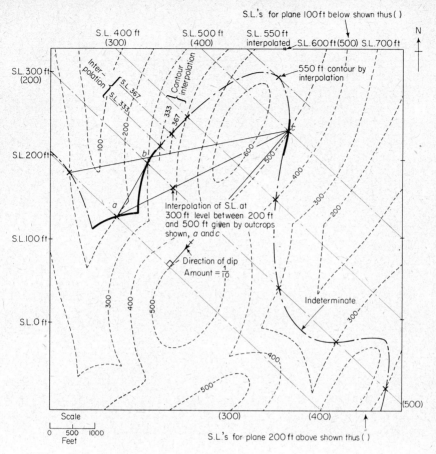

Figure. 4.24. Completion of outcrop (including 3-point problem for 'dip')
(By courtesy of T. Murby and Co.)

Question

1. The thickened portions of the lines shown on *Figure 4.24* indicate the outcrop of a bedding plane. Assuming this plane to have constant strike and dip, determine the amount and direction of dip and complete its outcrop throughout the area.

2. The student should complete the outcrops of two other planes, which are respectively 100 ft vertically below and 200 ft vertically above the one first drawn.

Answer

$$\tan \delta = \frac{100}{1,000} = \frac{1}{10}$$

$$\delta = 5 \text{ degrees } 44 \text{ minutes}$$

bearing S.W. approximately.

Note: Apparent 'dip' in direction *cb*

$$= \frac{200 \text{ ft vertical}}{2,400 \text{ ft horizontal}} = \tan^{-1}\frac{1}{12}$$

Note to Answer

The stratum contours of the uniformly dipping bedding plane may be drawn over the whole extent of the map after solution of the direction and amount of true dip from the 3 points of intersection *a*, *b* and *c* for the two portions of outcrop indicated. This gives the dip direction as S.W. and its amount as $\frac{1}{10}$.

135

Suggestions for sketching various cases of 'outcrops' are given as follows, so that the student may practice the principles already outlined. Some solutions can be found in Section 4.12 (6).

Compare the following:

(*a*) Bedding plane 'vertical'. Valley outcrops—strike crossing a valley at about 90 degrees.

(*b*) Bedding plane 'horizontal'. Valley outcrops—strike crossing a valley at about 90 degrees.

(*c*) Bedding plane (*i*) dipping upstream; (*ii*) dipping downstream. Note the changes in direction and angle of the V-shaped outcrop so formed when strata have dips as follows:

Figure 4.30

(*a*) Dip < slope of valley floor

(*b*) Dip = $\frac{1}{2}$ slope of valley floor

(*c*) Dip > slope of valley floor

Note: Solutions are given for case (*ii*) only, when stratum are dipping downstream. The student should attempt to draw the similar case (*i*) examples for himself.

(*d*) Folding. Synclinal and anticlinal outcrops (*Figure 4.30 d* and *e*)—plane uniformly sloping ground surface only.

Properties of and Variables affecting the Bed

As stated earlier, the simplest possible case of a stratum bed is one bounded by parallel plane bedding surfaces. This case seldom occurs in Nature except, perhaps, locally over comparatively short distances. In *Figure 4.25a* let α = angle of true dip; β = angle of ground slope in the true dip direction; w = width of outcrop projected on to the horizontal plane of the map; t = vertical thickness of the bed, i.e., the vertical distance between upper and lower bedding planes; T = true thickness of the bed, i.e., the perpendicular distance between upper and lower bedding planes.

The fundamental quantities of the bed are its 'true thickness', T, and 'true dip', α; the vertical thickness, t, depends on these and the width of outcrop, w, depends additionally on the angle of ground slope β. Given certain of these values, the remainder may be found by scaling from a graphical cross-section, or, more accurately, by trigonometrical calculation. It is a useful exercise to make scale plan and section diagrams of different cases and to calculate the required values of t, T and α from the width of a specific outcrop as measured on the plan, checking the result so obtained by measurement of stratum thickness on the cross-section view already drawn. From the plan view alone, vertical thickness and dip of a bed are often obtained by interpretation of strike line values for the outcrops of upper and lower bedding planes after noting their intersections with the ground contours. (*See* map example, *Figure 4.26.*)

Again, making practice sketches to show the effects of ground relief and stratum dip on outcrop shape for the simplest cases of a bed of constant thickness, is a useful exercise. From this it will soon be noted that outcrop width is most affected by change in the relative angle of ground slope and stratum dip (*Figure 4.25b*).

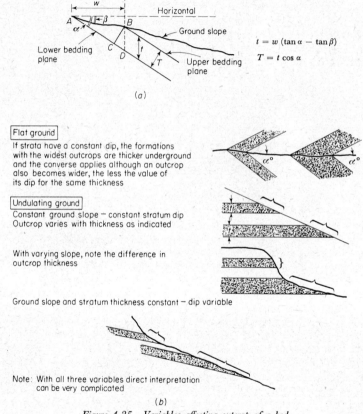

$$t = w (\tan \alpha - \tan \beta)$$

$$T = t \cos \alpha$$

(a)

Flat ground
If strata have a constant dip, the formations with the widest outcrops are thicker underground and the converse applies although an outcrop also becomes wider, the less the value of its dip for the same thickness

Undulating ground
Constant ground slope — constant stratum dip
Outcrop varies with thickness as indicated

With varying slope, note the difference in outcrop thickness

Ground slope and stratum thickness constant — dip variable

Note: With all three variables direct interpretation can be very complicated

(b)

Figure 4.25. Variables affecting outcrop of a bed

As previously mentioned and illustrated in *Figures 4.23* and *4.26*, the vertical thickness of a bed may be readily obtained from a contoured map by comparing the strike lines for the upper bedding plane with those of the lower bedding plane. In the simplest case where, say, the 1,200 ft S.L. of the upper plane coincides with the 1,050 ft S.L. of the lower bedding plane, the vertical distance between them is 150 ft and in general strike lines may be interpolated as necessary to obtain this sort of information.

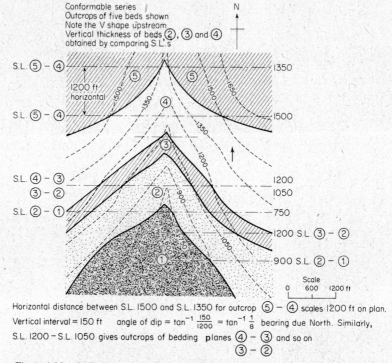

Horizontal distance between S.L. 1500 and S.L. 1350 for outcrop ⑤ – ④ scales 1200 ft on plan.

Vertical interval = 150 ft   angle of dip = $\tan^{-1}\frac{150}{1200} = \tan^{-1}\frac{1}{8}$ bearing due North. Similarly,

S.L. 1200 – S.L. 1050 gives outcrops of bedding planes ④ – ③ and so on
                                                               ③ – ②

Figure 4.26. Geological map showing simply dipping strata in valley head—dip upstream

(The 1,200 ft strike line for upper bedding surface of bed ③ coincides with 1,050 ft S.L. for lower bedding plane. Hence vertical thickness of bed ③ is 150 ft. 750 ft S.L. for lower bedding plane of bed ② coincides with 1,140 ft S.L. for upper bedding plane—by interpolation between 1,200 and 1,050 ft S.L.'s. Hence vertical thickness of bed ② is 290 ft. Similarly by interpolation the vertical thickness of bed ④ is 500 ft. Thicknesses of beds ① and ⑤ are indeterminate as their outcrops extend beyond the boundaries of map as shown. Careful labelling of S.L.'s for the different outcrops is necessary. S.L.'s only apparently coincide because the map is a horizontal projection of surface and underground relief.

Note: If the outcrops indicate a simple conformable series of beds, then one bedding plane will provide the answer for dip of all. Parallel S.L.'s indicate a conformable series; always check for this feature.)

Drawing Geological Sections

1. *Approximate method*—A convenient method of transferring surface points from a map when the section required is to the same scale as the plan, is by the use of a paper strip on which height points and the position of outcrop intersections along the section line are marked at relevant places. This strip is then transferred direct to the section plot below the map for sketching on it the ground profile and outcrop points. The section chosen should preferably be parallel to the direction of dip of the beds unless some other

special line is required or shows the geological structure more clearly. The vertical scale should, as far as possible, be the same (and on a 6 inch geological O.S. map section usually is the same) as the horizontal scale, because if too exaggerated a dip is shown it can be rather misleading. On 1 inch geological O.S. maps the vertical scale may be 3 times as great as the horizontal scale, but a section should never be as represented in much civil engineering survey work, where the vertical scale is often grossly exaggerated, e.g., 10 times the horizontal scale. For example, after inspecting the section on *Figure 4.27* try this approach with the *AB* line on *Figure 4.28*.

2. *General method*—The ground profile along the section line is first projected or plotted from points on the contours to their corresponding altitude lines on the section, at say 10 ft vertical intervals. The bedding planes are drawn by continuing S.L.'s across to the line of section and then projecting from the intersection points downwards or by plotting the positions of the intersections of these S.L.'s and the section line directly on to the sectional elevation below at the required heights. (*See Figure 4.23.*)

This procedure is more accurate than marking on the ground profile the outcrop positions and then drawing lines on the section at the 'correct dip' angles (which may well be apparent dips) since a small error in the profile as drawn will then cause a greater error in the thickness and position of the beds. The S.L. method also checks the accuracy of the sectional profile, as the outcrop positions on it can then be projected upwards and compared to the original plan view; beds are plotted in reverse order to age, topmost first. They should then be shaded or coloured as on the plan view and their true thickness measured or calculated.

Summary

1. *To determine the dip of a bed from its outcrop on a map*

(*a*) Draw a strike line for one bedding surface—upper or lower. Note its value.

(*b*) Draw a second S.L. for the same bedding surface. Note its value.

(*c*) Measure at right angles the horizontal distance between them and calculate

$$\frac{\text{Vertical interval}}{\text{Horizontal distance}} = \tan^{-1} \text{ (Angle of true dip)}$$

2. *To determine the vertical thickness of a bed similarly*

(*a*) Draw strike lines for one bedding surface—preferably the upper. Note their values.

(*b*) Draw strike lines for the other bedding surface. Note their values.

(*c*) Compare the upper and lower sets of S.L.'s—looking particularly for coincidence—and deduce the difference in level between any corresponding pair of these particular S.L.'s. (*See Figure 4.26.*)

3. *To determine the vertical displacement due to a fault*

(*a*) Draw strike lines for any convenient bedding surface on one side of the fault. Note their specific values.

(*b*) Draw strike lines for the same bedding surface on the other side of the fault. Note their specific values.

(*c*) Compare the two sets of S.L.'s. Look particularly for coincidence of two particular S.L.'s where they abut on either side of the fault plane and compare their values for the 'throw of the fault', otherwise interpolate pairs of S.L.'s as required. (*See* Chap. 5, Example *Figure 5.5*.)

The ability to read and understand geological maps is an essential feature in the training of civil engineers today and may be needed at any time, whether in a personal site investigation or in the interpretation of reports from an expert geologist.

Apart from the examples given in this Chapter (with acknowledgement to Platt's *Elementary Exercises on Geological Maps*), a series of graded map exercises are included in Platt's Exercises and provide practice in interpreting a wide variety of underground geological structures from the forms of strata outcrops shown on plan views.

4.11. EXERCISES

1. Describe briefly, with the aid of sketches, the following geological phenomena, giving some well-known examples of each:

(*a*) Dip and strike of beds outcropping in the form of a 'Scarp'.

(*b*) Anticline and syncline; monocline and isocline.

(*c*) Normal fault and tear fault.

(*d*) Unconformity, including the series of changes which lead to any typical case you have illustrated.

2. (*a*) Sketch and describe each of the following igneous features:
(*i*) Boss and batholith; (*ii*) dyke and sill; (*iii*) volcanic plug or neck, listing any well-known examples with which you are familiar.

(*b*) (*i*) Discuss how their mode of occurrence influences the texture and type of igneous rock produced in each case above, naming some of the more important rocks so formed. (*ii*) What metamorphic effects are associated with such igneous phenomena? Give various examples for the features mentioned in (*a*) above.

3. (*a*) Describe the commonly occurring types of fault and illustrate your answer with diagrams showing their effects on the outcrops of a series of simply dipping strata.

(*b*) (*i*) Explain the meaning of the terms, *throw, heave, shift* and *hade* with reference to either a normal or reversed fault, by sketching in section a 'normal strike fault' hading in the same direction as the dip of 'strata', which include a 'sill' and are overlain by an 'unconformity'.

(*ii*) Illustrate also in plan and section the form of a 'pitching asymmetrical anticline' and distinguish between the relative value of 'bedding', 'jointing',

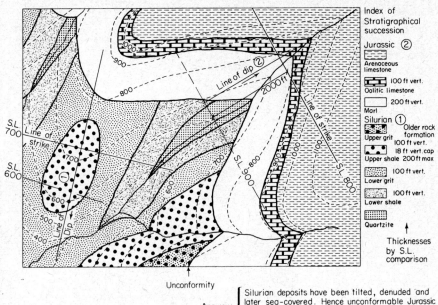

Index of
Stratigraphical
succession

Jurassic ②

Arenaceous
limestone

100 ft vert.
Oolitic limestone

200 ft vert.
Marl

Silurian ①
Older rock
Upper grit formation
100 ft vert.
18 ft vert. cap
Upper shale 200 ft max

100 ft vert.
Lower grit

100 ft vert.
Lower shale

Quartzite

Thicknesses
by S.L.
comparison

Unconformity

Answer: { Silurian deposits have been tilted, denuded and later sea-covered. Hence unconformable Jurassic marine sediments, themselves later tilted, to produce an angular unconformity

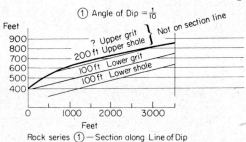

① Angle of Dip = $\frac{1}{10}$

Feet

? Upper grit } Not on section line
200 ft Upper shale
100 ft Lower grit
100 ft Lower shale

Rock series ① — Section along Line of Dip

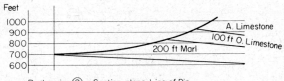

② Angle of Dip = $\frac{100}{2000} = \frac{1}{20}$

Feet

A. Limestone
100 ft O. Limestone
200 ft Marl

Not to scale

Rock series ② — Section along Line of Dip

Figure 4.27. Unconformity
(Map by courtesy of T. Murby and Co.)

'cleavage' and 'dip' in relation to the quarrying or mining of rock strata.

4. (a) What is meant by the following geological terms: *Strike; true dip; apparent dip; escarpment; outlier* and *inlier*.

Explain in particular the origin and nature of the latter three phenomena.

(b) A, B, C, D are four boreholes situated respectively at the N.W., N.E., S.E. and S.W. corners of a square with sides 1 mile long. Ground levels at A, B, C and D are respectively 1,260 ft O.D., 1,470 ft O.D., 1,250 ft O.D. and 1,060 ft O.D. In the boreholes A, B and D, the top of a coal seam was found at 1,100 ft O.D., 860 ft O.D. and 660 ft O.D., respectively.

Assuming a uniform dip to the coal seam and uniform gradients on the ground surface, draw the points on plan to scale of 1in = 1,000 ft, determine the direction and amount of dip of the seam and the depth of the coal below ground surface at C.

5. Draw sketches to illustrate the causes of: (a) two faults aligned to form a 'ridge faulting'; (b) a denuded 'dome'; (c) a 'sill' with a feeder 'dyke' which have been (subsequent to their formation) tilted with their surrounding strata and overlain unconformably by a horizontally bedded rock series.

6. (a) Show diagrammatically how the width of an 'outcrop' for a stratum of constant thickness can be affected by the relative dip of the stratum and slope of the ground surface.

(b) Draw sketches to illustrate the contours of a rising river valley and superimpose the outcrops of strata which are (i) dipping upstream; (ii) dipping downstream. In both cases, show examples with the dip of the strata both greater and less than the slope of the ground surface.

7. Write an account of earthquakes and their effects, both as apparent on the earth's surface and also with regard to the information deduced about the interior structure of the earth, from a study of their various shock wave emanations.

8. Determine completely the geological structure of the rock outcrops indicated on the plan area shown (*Figure 4.27*), and draw sections as required to confirm your conclusions. (A complete analysis by S.L.'s for all beds should then be attempted.)

The horizontal scale of any section can be the same as that of the map.

4.12. PRACTICAL WORK—SOME SUGGESTIONS

1. Students should normally relate their practical work to the study of geological maps and may best begin with exercises on 3-point problems and maps, similar to those given in this text (Section 4.10).

The study of block models and relief models with sections, to show the inter-relation of outcrop with topographic form and the use of a surveyor's compass and clinometer for measuring actual depths of outcrops in the field should also be undertaken. If the student can visit a suitable cliff, quarry or site during his field studies and relate his observations to the bedding and jointing, dip and strike, of the facial rocks exposed (*Figures 4.1, 4.9 and 4.16*), this should help him considerably.

Actual geological survey maps (*see* references and cross-sections listed in Section 2.8) when used for field work, and drawing sections with as little vertical exaggeration as absolutely necessary, greatly increase the value of these exercises and help in acquiring interpretive knowledge of plan views.

Introductory work should be attempted with 'folding', 'faulting', 'unconformity' and 'igneous intrusions' on geological O.S. maps, as being vital to a clear understanding of the fundamental principles to be applied to engineering and map problems.

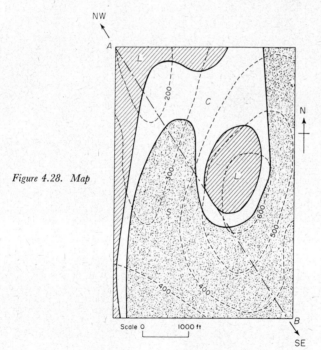

Figure 4.28. Map

The use of local 1 inch geological O.S. maps, together with the appropriate 1 inch Sheet Memoir or Regional Handbook and the 1 inch topographic map, will all provide useful practice.

2. (*a*) Obtain an outline geological map of the British Isles (*see Figure 1.4*) and colour it as nearly as possible with the standard shades used on the 25 miles to 1 in sheet, noting also the standard system of symbols for various rock strata and types (as shown on 1 in maps). Learn in passing, how Great

Britain and particularly the Midlands, south and south-east England are divided into a number of natural banded regions reading from N.W. to S.E.; each peculiar to a certain geological system.

3. *Geological Features*

The following list of postcards, etc., from the London Geological Museum, I.G.S., is especially relevant to the study of geological structures.*

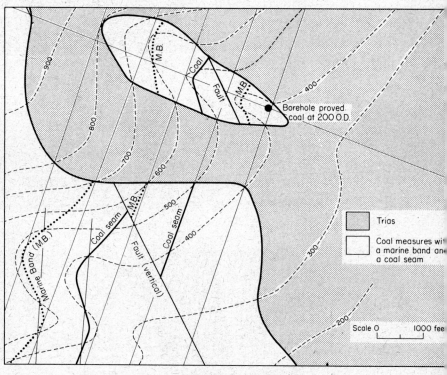

Figure 4.29. Map

Vulcanism. Scenery of Igneous rocks: MNL 522, MNL 514 to B 928; CP 1, CP 54, C 3655, 3661, 3895; A 6492; B 474.

Scenery of Sedimentary rocks: MNL 525; MN 2612; A 7607, A 7542; CP 27, 40, 55, 59.

Reconstructions: Dioramas, MN 2613; MNL 519, 459, 5258.

* See the '*Short Guide to the London Geological Museum*', and Classified Geological Photographs, H.M.S.O.

4.12. PRACTICAL WORK—SOME SUGGESTIONS

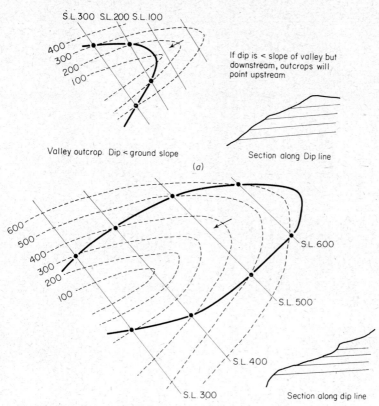

If dip is < slope of valley but downstream, outcrops will point upstream

Valley outcrop. Dip < ground slope

Section along Dip line

(a)

Valley outcrop. Steep ground slope – small dip value downstream

Section along dip line

(b)

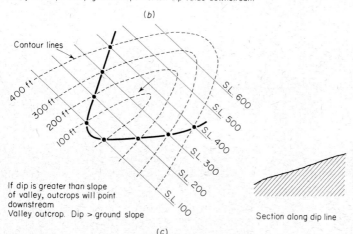

Contour lines

If dip is greater than slope of valley, outcrops will point downstream

Valley outcrop. Dip > ground slope

Section along dip line

(c)

Figure 4.30 a-c (continued overleaf)

145

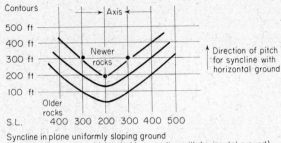

Syncline in plane uniformly sloping ground
(Similar effect produced by pitching syncline with horizontal ground)

(d)

Folding of strata – outcropping on a horizontal or simply sloping ground surface

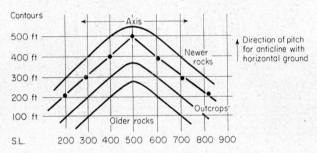

Anticline in plane uniformly sloping ground
(Similar effect produced by pitching anticline with horizontal ground)

(e)

Figure 4.30. Suggestions for sketching outcrops

Chalk and Limestone Country: A 5414; CT 19, 53.

Structural Geology: NI 283, 284; A 5769, 9984, 10044; MN 17602, 17604, 17605, 20499, 20500, 20501; CP 2, 3, 6–8, 12, 15, 30, 32–34, 36, 43–46, 56; MNL 501, 513, 523, 533.

Photographic Booklet No. 2—10 views of rock structures.

4. *Interpretation of Geological Maps*

In a small area of simply dipping strata (sedimentary) the *outcrops* of three *formations* have been traced and their distributions plotted on the accompanying map (*Figure 4.28*). Determine accurately by means of labelled *strike lines*:

(*a*) The direction of Dip.
(*b*) The amount (in degrees) of the Dip.
(*c*) The 'succession' of strata.
(*d*) The true 'thickness' of the *unshaded* formation.

146

(*e*) Draw a section along the line of true dip through the summit of the hill.

(Try to obtain a three-dimensional mind picture of the area.)
Insert your results on the map by dip arrows and show a 'Geological column' in the margin.

5. On the map shown (*Figure 4.29*) indicate by shadings, the areas under which the coal seam is present.

6. Solutions to suggestions for sketching various cases of outcrops. (*Figure 4.30 a-e.*)

REFERENCES AND BIBLIOGRAPHY

Billings, M. P. (1959). *Structural Geology*. 2nd edn. London and New York; Prentice-Hall

Blyth, F. G. H. (1965). *Geological Maps and their Interpretation*. London; Edward Arnold

Daly, R. A. (1938). *Igneous Rocks and the Depths of the Earth*. 2nd edn. New York; McGraw-Hill

Davison, C. (1936). *Great Earthquakes*. London; T. Murby

Dowrick, D. J. (1968). 'Current Practice in Earthquake Resistant Design.' *Proc. Instn. civ. Engrs.* **40,** July

Dunbar, C. O. and Rodgers, J. (1951). *Principles of Stratigraphy*. New York; Wiley

Eardley, A. J. (1962). *Structural Geology of N. America*. New York; Harper and Row

Eastwood, T. Compiler (1965). Stanfords Geological Atlas of Great Britain. London; G. Philip and Sons

Ginoux, M. (1955). *Stratigraphical Geology*. U.S.A.; W. H. Freeman

Himus, G. W. and Sweeting, G. S. (1955). *The Elements of Field Geology*. 2nd edn. London. Univ. Tutorial Press

Hodgson, J. H. (1964). *Earthquakes and Earth Structure*. London; Prentice-Hall

Matuzawa, T. (1964). *Study of Earthquakes*. Tokyo; Maruzen Co.

Platt, J. I. (1945). *A Series of Elementary Exercises on Geological Maps*. 3rd edn. London; T. Murby

Platt, J. I. (1964). *Selected Exercises upon Geological Maps*. 2nd edn. London; Allen and Unwin

Platt, J. I. and Challinor, J. (1949). *Simple Geological Structures*. 3rd edn. London; T. Murby

Rayner, D. H. (1967). *The Stratigraphy of the British Isles*. Cambridge University Press

Read, H. H. (1957). *The Granite Controversy*. London; Allen and Unwin

Rigley, J. E. (1935). *The Tertiary Volcanic Districts*. (3rd edn. 1961.) British Regional Geology Handbook—Scotland. London; H.M.S.O.

Shand, S. J. (1959). *Eruptive Rocks*. London; T. Murby

Smithson, F. (1950). Patterns for a Series of 12 Block Models Representing Geological Structures

Stamp, L. Dudley (1959). *Britain's Structure and Scenery*. 5th edn. London; Collins

U.S. Geological Survey (1932). Geological Map of the U.S.A.

Zumberge, J. H. (1963). Elements of Geology. 2nd edn. New York; Wiley

CHAPTER FIVE

ENGINEERING GEOLOGY AND GEOLOGICAL MAPS

5.1. INTRODUCTION

The elements and principles of geology will now receive increasing application in the practical engineering sense; thus, in this chapter a review of mapping with a bias towards simple engineering problems is attempted, so that the student may consolidate his theoretical knowledge before various aspects of engineering geology are discussed in some detail.

It is possible by the process of geological map reading, to infer from the forms and types of outcrops as they intersect the surface topography, not only the underground geological structure but also the nature of particular rocks; it is often possible also to deduce probable rock properties, even at some depth below the ground. However, field work to support and confirm any preliminary conclusions is usually necessary, although this may be minimized by intelligent use of map information shown. It is through a suitable combination of these two factors that recommendations for constructional works or other relevant projects are given and so reference will first be made to map data, etc. available for such purposes.

5.2. PRODUCTION AND USE OF GEOLOGICAL MAPS

Maps and other Data obtainable from H.M. Geological Survey I.G.S.

The standard geological map is published in the 1 inch to 1 mile editions and is based on the 1 inch topographical O.S. sheets, which should be familiar to most readers. The whole of Great Britain was surveyed and hand-coloured sheets of the 'Old Series' produced from this between 1840 and 1920. Subsequent revision has provided the New Series colour printed maps (with standard symbols and type), issued in two editions, Solid and Drift, the latter wherever required.

On the Solid maps, which are very useful for mining and tunnelling problems, the geology is shown as the outcrops would appear if their mantle of soil and vegetation were stripped away. The Drift maps, which are most important when questions involving soils arise, show all superficial deposits (i.e., river gravels, alluvium, residual soils, boulder clay, etc.) with the Solid rock geology completely obscured, except where it outcrops through any overlying Drift.

Although the Old Series geological maps were commenced over 120 years ago and geological survey then was based on the existing 1 inch O.S. editions, the preparation of the 'New Series' has benefited from the availability of 6 inch and 25 inch to the mile O.S. maps. The 6 inch O.S. map is now the usual basis for geological survey, with assistance from the 25 inch maps (if available) in areas of great complexity.

The 6 inch geological maps are not generally published (but may be seen at the Geological Survey/Museum Library, London), the available 1 inch maps being a reduction from these by the Geological Survey, although printed copies of large-scale maps may be issued on request—especially for coalfield areas and industrial or residential/town areas. Regionally, use is best made of the $\frac{1}{4}$ inch to the mile maps, showing the geology in less detail of whole counties and excellently supplemented by their Regional Handbooks, which describe the geological features, history, topography, structure, water supply and economic products of the particular region. (An I.G.S. index map of 18 regions is available free.) The 1 inch Sheet Memoirs giving full explanation of the above features and issued for each 1 inch map sheet, describe the particular area shown in very complete detail. There are also the very useful supplementary booklets on water supply (Memoirs containing records of wells and underground water sources for English counties) and other topics, such as coalfields, economic and mineral resources as fully catalogued in H.M.S.O. Sectional List No. 45; these include also the various District Memoirs, Special Memoirs and Reports, etc. A 'List of Geological Survey maps' is supplied free by the I.G.S., O.S. and agents on request.*

It is best to confirm from the Director of the I.G.S., London, whether any unpublished information such as aerial photographic mosaics, separate area photographs, borehole records, etc., exist for a particular area.

Much original manuscript-form material is also obtainable.

References to individual map sheets and memoirs are given in this book where appropriate and the student should familiarize himself with the symbols used on the various 1 inch maps listed.

Other countries like the U.S.A., Canada, Australia, those of Continental Europe, etc., similarly publish a comprehensive series of O.S. and geological type maps; the necessary information is obtainable from their relevant Government Survey Departments.

Use of O.S. and Geological Maps, etc.

The value of maps, as giving a most precise pictorial summary of the topography and geology of an area (and especially when combined with

---

* In the U.S.A., British Government publications are obtainable from the British Information Services, 845 Third Avenue, New York 22, N.Y., and catalogues may be inspected at British Consular Offices through the country, as may be done also at major British Consulates in all parts of the world. There are Overseas Agents in 36 countries, Government Bookshops and Agents in 40 cities and towns of the U.K.

B.S.I. offices are situated throughout Canada and in India also.

other relevant data), should be evident, together with their consequent importance to the geologist, civil engineer and builder, prospector, mineralogist, miner and others engaged in the selection of sites for the works of man. For example, to investigate preliminary schemes for the exploitation of mineral and economic resources by mining and quarrying, or for winning limestones, brick clays, building stones, road metals, concrete aggregates: in producing suitable proposals for the building of dams, tunnels, roads, railways, etc.; in the prediction of water table levels and the quantities/quality of anticipated underground/surface water supplies and/or catchments —all such projects are first better explored on O.S. topographical and geological maps. Wherever the geological structure, type of rock and its predicted location at depth are required, the geological map, supplemented by carefully planned field work and borehole records, enables the necessary recommendations to be made. Full information on the objects, scope, data required and means of carrying out such investigations is given in *Civil Engineering Code of Practice No. 1* (CP 2001), *Site Investigations* (Appendices F and H especially), issued by the Institution of Civil Engineers and the B.S.I.; this is the practising engineer's reference handbook and guide to the suitability and characteristics of sites or regions for projected works— it summarizes in the most convenient form virtually all the possible factors to be considered, together with the sources and availability of any relevant information required.*

5.3. GEOLOGY AND SOILS

Weathered products of the denudation processes which collect *in situ*, rather than those which are readily and continuously removed by such agents as wind, water, etc., are the basic materials from and upon which a surface soil layer is developed. Thus, top soil, sub-soil and parent rock (*Figure 5.1*) are definitely interrelated in their appearance and properties when the soil is a deposit of this residual type. When a transported cover of drift material, such as glacial clay, overlies the natural bedrock, however, correspondence of their properties cannot be so marked, although in course of time even a transported superficial deposit will possibly inherit some of the solid bedrock's characteristics; this is especially so, where these latter properties emanate from projecting outcrops and are due to substances leached away from these as the weathered products of rainwater, to become absorbed in the neighbouring (albeit originally foreign) top soil.

To the agriculturalist and most geologists, the word 'soil' means 'pedalogical soil' or that variable mixture of humus and mineralogical rock waste which contains milliards of living organisms, whether due to plant

---

* Extracts from this code are quoted where relevant hereafter (*see also* CP 2002, 2003, CP 2004 and 2006).

and animal, or to insect and minute bacteriological life. The resultant geological outcome in any specific region is chiefly controlled by the prevailing climate under which a particular soil type exists (*Geomorphology*).

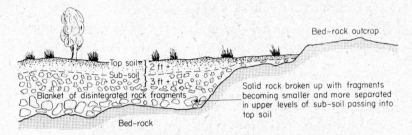

Figure 5.1. Soil and rock

Two typical examples. Residual deposits: (a) Parent bedrock of sandstone: soil—light sandy variety. (b) Parent bedrock of granite: soil—kaolin or china clay variety
Transported debris: Glacial drift forming boulder clay topsoil, over any type of bedrock

Thus, according to the cyclic state of equilibrium, or disequilibrium with time, in any one region, as between soil-removing and soil-forming processes, e.g., topographic, climatic, biological, chemical, and whether these natural agencies cause degradation into the bedrock or deposition and anchorage of loose matter in existing vegetation, therefore various consistent soil types predominate, which are the integrated and complex results of the above variables. In most regions a continuous layer of soil exists with its protective surface of vegetation and despite the leaching action of rainwater or the effects of wind and ground water movement sufficient humus remains behind in the top soil as the raw material for plant growth which, in turn, acts as a support for animal life. However, a broad distinction can be made between lands under temperate climatic conditions, where weathering usually takes place mainly *in situ* and soil types are often closely related in characteristics to their bedrock, and those lands subjected to extreme tropical or arctic climatic conditions, when many of the soils produced may be totally unrelated to the regional solid geology.

Engineering Aspects

Although geologists classify all sedimentary deposits, whether 'solid' or 'drift' as rocks, for practical purposes engineers divide such deposits into the two major groups, 'rocks' and 'soils'. The term soil is thus used with reference to all relatively loose, part-compacted, or uncemented softer materials, and to the engineer, rock means only the hard, rigid, igneous, metamorphic and well cemented sedimentary types. The relationship between the broad engineering classification of soils and rocks and the

151

geologist's or agriculturalist's meaning of soil applied in the pedological* sense, are shown in *Figure 5.2.*

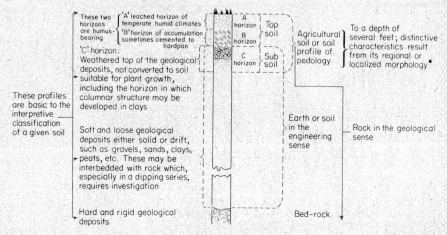

Figure 5.2. The relation between pedological soil and soil in the engineering sense—Cross-sectional profile

(Note: In Great Britain, natural seasonal changes in moisture content, affect soils to a depth of over 5 ft, which is therefore the minimum depth for site exploration related to shallow founded structures such as roads and airfield runways. Frost seldom penetrates to a depth greater than 2 ft and is not usually a critical factor in engineering operations.

Extreme climatic conditions such as occur in parts of Canada and the U.S.S.R. between summer and winter cause variations in ground moisture to depths of 15 ft and can effect soil weathering to a very considerable extent).

*Morphology: the integrated effects of variables such as climate, bedrock, vegetations, bacteria, land topography and time (after Civil Engineering Code, CP 2001).

Soil types include any materials derived as detrital sediment such as gravel, sands, silts and clays; as organic deposits like peat, or chemical deposits like chalk; as *in situ* residual weathered matter such as certain clays and tropical 'laterites'; as calcareous shelly accumulations of sand, etc.; and even, as explained in Chap. 2, material of volcanic origin known as pyroclastic sediments; thus previous reference has already been made to most sedimentary deposits and their variegated origins. Soils are produced in nature by such a variety of processes that a most diverse and complex material can result, as happens with the clays on which many engineering structures are built.

For engineering identification purposes, only two important physical characteristics of most soils need to be considered.

---

* Pedological earth thus means the 'top soil' and part 'sub-soil' containing mineral and vegetable matter which, after long subjection to bacteriological action, forms the organic 'humus' necessary for plant growth and lends its darker, even black, colour to the visible soil surface and the layers immediately beneath. Pedological earth, when it is very deep, may well extend into the horizons defined as soil in the engineering sense, as with 'laterite' in tropical regions and 'hardpan' in cooler continental climates, but in Great Britain the two meanings are always distinct.

Table 5.1. General Basis for Field Identification and Classification of Soils

(From *Code of Practice* CP 2001 reproduced by permission of the B.I. Institute, 2 Park Street, London W.1, from whom copies of the complete Code of Practice may be obtained)

		Size and nature of particles		Composite types 3
		Principal soil types 1	2	
		Types	*Field identification*	
Coarse grained, non-cohesive		Boulders Cobbles	Larger than 8 in in diameter Mostly between 8 in and 3 in	Boulder gravels
				Hoggin
		Gravels	Mostly between 3 in and No. 7 B.S. sieve	Sandy gravels
	Uniform	Sands	Composed of particles mostly between No. 7 and 200 B.S. sieves, and visible to the naked eye. Very little or no cohesion when dry. Sands may be classified as uniform or well graded according to the distribution of particle size. Uniform sands may be divided into coarse sands between Nos. 7 and 25 B.S. sieves, medium sands between Nos. 25 and 72 B.S. sieves and fine sands between Nos. 72 and 200 B.S. sieves.	Silty sands Micaceous sands Lateritic sands Clayey sands
	Graded			
Fine grained, cohesive	Low plasticity	Silts	Particles mostly passing No. 200 B.S. sieve. Particles mostly invisible or barely visible to the naked eye. Some plasticity and exhibits marked dilatancy. Dries moderately quickly and can be dusted off the fingers. Dry lumps possess cohesion but can be powdered easily in the fingers.	Loams Clayey silts Organic silts Micaceous silts
	Medium plasticity	Clays	Dry lumps can be broken but not powdered. They also disintegrate under water. Smooth touch and plastic, no dilatancy. Sticks to the fingers and dries slowly. Shrinks appreciably on drying, usually showing cracks. Lean and fat clays show those properties to a moderate and high degree respectively.	Boulder clays Sandy clays Silty clays Marls Organic clays Lateritic clays
	High plasticity			
Organic	Peats		Fibrous organic material, usually brown or black in colour.	Sandy, silty or clayey peats

Note. The principal soil types in the above table usually occur in nature as siliceous sands and silts and as alumino-siliceous clays, but varieties very different chemically and mineralogically also occur. These may give rise to peculiar mechanical and chemical characteristics which, from the engineering standpoint, may be of sufficient importance to require special consideration. The following are examples:

Lateritic weathering may give rise to deposits with unusually low silica contents, which are either gravels or clays; but intermediate grades are rare.

Volcanic ash may give rise to deposits of very variable composition which may come under any of the principal soil types.

Deposits of sand grade may be composed of calcareous material (e.g., shell sand, coral sand) or may contain considerable proportions of mica (where grain shape is important) or glauconite (where softness of individual grains is important).

Deposits of silt and clay grade may contain a large proportion of organic matter (organic silts, clays, or muds) and clays may be calcareous (marls).

Loam is a sand, silt and clay mixture with these three types as approximately equal proportions of the whole il mass.

1. The predominant particle size and nature of these or other particles forming a soil type.

2. The density* (and in the case of clays, also the 'plasticity') of a soil which results from the *in situ* arrangement of its constituent particles.

The above, in conjunction with a number of simple Field Identification Tests, including visual and textural examination, enable engineers to classify soils into the six main types shown in Table 5.1, with a principal distinction being made between non-cohesive, cohesive and organic soils.

Particle size limits are based on those given in Table 2.2 and mechanically formed granular-type soils (e.g., boulders, gravels, sands) being non-cohesive, are classified mainly by their range of particle sizes. On the other hand, cohesive soils are classified mainly on the basis of their characteristic property of 'plasticity'. Typical particle size distribution curves for various soils are shown in *Figure 5.3.**

These divisions may appear somewhat arbitrary but work well in practice because they represent important changes in engineering properties of the various soil types listed, such as their drainage characteristics. The whole basis of soil classification from the engineering standpoint is amply explained in a D.S.I.R. publication,† where the 'Extended Casagrande System of Soil Classification' is detailed with the standard letter symbols used for soil groups, although Table 5.1 is largely self-explanatory.

To summarize briefly, the usual engineering description of a soil type would be couched in the following terms and group symbols attached, if possible:

Type and grading of particles; density and/or structure; colour and odour, if present; e.g., Thus 'a stiff, fissured, brown boulder clay', 'a coarse, slightly cemented, micaceous sand'.

The science of Soil Mechanics is Applied Geology in the sense of quantitative investigations and measurement of the engineering properties of soils in relation to proposed and actual constructions, for which, investigations cover field tests *in situ* and/or laboratory measurements on soil samples obtained from site boreholes.

5.4. GEOLOGICAL MAPPING IN THE FIELD

The problem involved here is that of plotting as far and as accurately as possible by intelligent interpretation of all available evidence, the actual 'outcrops' of various geological formations where exposed on the ground surface—or at least, nearly so—within the proposed map area. (Note: This is not quite the same exercise as that of the simpler 'Completion of

---

* Soil classification procedure is fully described in B.S. 1377 and sampling methods in CP 2001. Bulk density is related also to natural moisture content, which is variable.

† *Soil Mechanics for Road Engineers*, an authoritative and comprehensive text on soils and soil engineering.

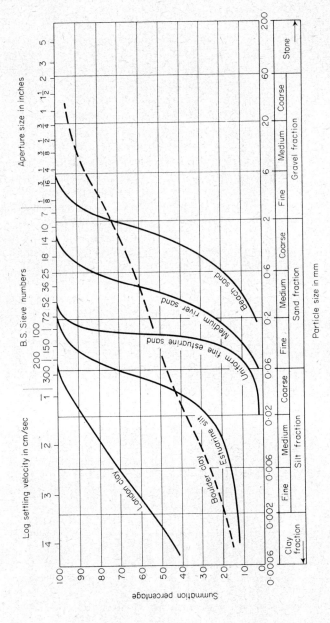

Figure 5.3. Typical particle size-distribution curves

(From Civil Engineering Code CP 2001, by courtesy of The British Standards Institution)

outcrop' on a mapped ground plan through the intersection of S.L.'s obtained from a known geological structure with given surface topographic contours as described in Chap. 4.)

Many difficulties due to the blanket of surface vegetation, drift (or superficial deposits) and top soil occur, and often a small part only of the true 'solid geology' is visible. Nevertheless, the fuller interpretation of any preliminary ideas concerning the geological structure/stratification is largely made from observations of beds seen in cliffs, crags, streams, wells, quarries, mines, cuttings for road, rail and canal works, supplemented where necessary by boreholes, trial pits and trenches. A great deal of personal and highly experienced interpolation and extrapolation is required by the surveyor, as is the need for producing both 'drift' and 'solid' maps in most regions of countries like Great Britain. Dip and strike predicted from one point of an 'outcrop' will often be confirmed elsewhere across country at a further exposure of the same rock type; however, 'dip' is seldom constant and 'faulting' occurs often in practice so that such deductions must of necessity be localized. The topography and weathering effects, and the position of the water table, shown by lines of springs, lakes, drainage etc., are all most revealing in substantiating conclusions regarding the interaction of the ground surface with the underground structure and every possible type of field observation may be useful. The colour of top soil, type of vegetation, spoil from animal burrows and direct borings through 'drift' by hand auger to depths of 10 or 20 ft—all assist in establishing the solid underlying surface. The information thus obtained is plotted on topographic maps using standard symbols of CP 2001, APP. C, together with a standard colour system for different rock types, if required, and existing Ordnance Survey 6 inch to the mile sheets are invaluable in aiding the personal ideas and conclusions reached by investigators, when engaged in site investigations for engineering works.

Aerial survey is today providing an increasingly important and economic means of producing topographic and geological maps of large regions, by the skilled interpretation of photographs (*Figure 5.4*); this applies especially to the discovery and exploitation of mineral resources in undeveloped and unmapped countries of the world (in South America, Africa and the Middle East, for example), where the terrain is extremely inaccessible and too difficult for conventional ground survey methods. Here, the varied shades of vegetation and those of the different soil/rock types also, as well as the evidence of drainage patterns and differential weathering in forming distinctive topographic features of a landscape, are the main means of expert study and help the geological surveyor by his interpretation of stereoscopic pairs of aerial photographs to produce ultimately the required maps or recognize the presence of an economically exploitable and viable source of mineral wealth. The Photogeological Division of the I.G.S. (*see* Whittle and Hepworth, 1967) and specialist firms, such as Hunting Surveys and B.K.S. Ltd. in England, who are engaged in this type of work, are usually willing to supply the student, on request, with relevant material and information in the form of pictorial brochures and brief explanatory leaflets of their activities in fields of operation throughout the world.

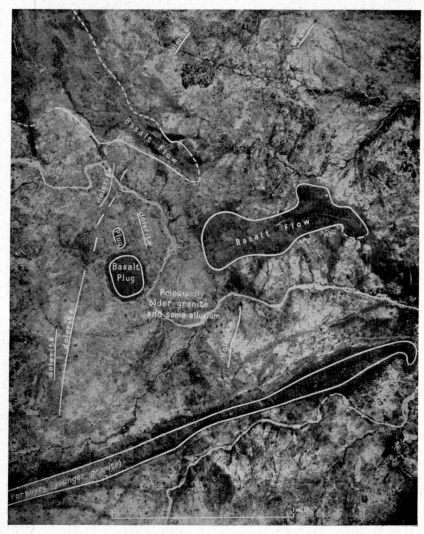

Figure 5.4. Example of geological survey from the air

(By courtesy of Hunting Aerosurveys Ltd.)

The evidence collected has to be fitted together after the manner of a jigsaw puzzle, then checked by every possible means. No map is ever complete in every detail but may be accepted as a reasonable interpretation of the available evidence and it is always subject to amendment in the light of new information, generally obtained from engineering (e.g., cuttings, canal, tunnel or mining) works. The O.S. and Geological Survey Departments of Great Britain are constantly revising their various maps and publications. Old Series 1 inch maps are now long out of date and even the earlier maps of the New Series list are already in course of revision.

5.5. APPLICATION OF MAP DATA FOR ENGINEERING PROBLEMS

Description of Map Evidence

To write a satisfactory description of a geological map requires that the evidence is dealt with carefully and systematically in a suitable order, outlined as follows; but of course all the points listed do not necessarily apply to every map. The systematic approach given here should be followed by the student, but practice is required before real proficiency can be gained in ensuring a complete interpretation of the information given pictorially of any district.

For each rock series or group in the area shown; study:

1. *The succession of beds*—Tabulate these according to their age from the surface downwards, grouping them into systems, and formulate a general idea of their geological structure. Give the values of vertical thicknesses and label the newest and oldest beds.

2. *The detailed structure*—Starting with the newest system, note down the following facts:

(*a*) *Dips and thicknesses of beds*—i.e., note the direction and amount of 'true dip', including any local variations in this and similarly in the thicknesses of beds.

(*b*) *Folding*—Note which beds are folded, deduce the types of fold present with values of 'dips' on their 'limbs' with the direction of their longitudinal axes, and try to estimate also the age of any folds.

(*c*) *Faulting*—Deal with each 'fault' separately and state the types of fault present; give for each its hade and line of direction, i.e., whether generally that of the strike or dip of beds; indicate the downthrow side and the values of heave and throw; estimate the age of faulting and state exactly which beds are involved. Comment also on the relation between the various faults, if there are more than one; i.e., whether criss-crossing a region erratically, or having a more parallel arrangement as a definite connected series.

(*d*) *The inter-relation* of any 'folding and faulting', if apparent, should be noted also.

3. *The relation of rock groups or series*—The position and nature of 'unconformities', alteration of 'strike' line directions, etc., should be listed for each series of beds.

4. *Any igneous phenomena*—State the forms of 'intrusion or extrusion' evident on the map; i.e., laccolith, dyke, sill, flow, sheet, etc.; note also their extent, both in area and depth, estimated age, and also the 'sedimentary rock beds' affected locally in shape or through metamorphism, etc., by each igneous occurrence, with the formation, shape, size, depth, etc., of any 'aureoles' present.

5. *Any special features*—Note such features as drift, alluvium, raised beaches, terraces, evidence of glacial deposits and glacial action, etc., with their extent and total blanketing effect on the solid geology.

6. *The topography*—Study the existing ground surface carefully for weathering and denudation features, drainage pattern, etc., and deduce their interrelation with the geological structure of the mapped area.

7. *Summary*—Finally, summarize as far as possible, the geology of the whole area by historical correlation of the above data and your own conclusions derived therefrom. Also include in your summary particular mention of rocks or minerals, etc., which have some economic value (e.g., as building stone or road materials) and note any other factors such as the presence, if any, of water supplies from underground catchments and like features. Give a brief statement of the location of these rocks, etc., in relation to that of the region mapped with this area also correlated to its surrounding districts and orientated in the wider sense of National Grid references or the true north direction.

Outcrop Trends—A Summary

1. A rise in ground level causes the 'trend' (or general direction) of and outcrop to deviate in a direction opposite to the direction of its dip.

2. A fall in ground level causes the trend of an outcrop to be diverted towards the direction of its dip. Whenever an outcrop turns towards a lower contour there must be some dip in that direction.

Valley outcrops—Where an outcrop crosses a valley, it forms a 'V' shape in plan and usually, the 'V' of outcrops points downstream, the dip of beds is likewise downstream—a most useful rule to remember in map reading.

Folding and variation of dip

In proceeding outwards from an axis of folding,

the structure is 'synclinal' when the outcrops of successively 'older' beds are met in order as distance from the 'axis' increases, and the reverse applies to an anticline.

Unconformities

1. Show a marked discordance in the 'strike of outcrops'.
2. Show a marked and sudden change in the 'amounts of dip' over a large area.
3. Show gradual overlapping of 'outcrops of older beds' by those of 'younger beds'.
4. Show a gap in the chronological order of strata.
5. Show igneous intrusions like dykes, which traverse some outcrops and end abruptly at other outcrops, or 'fault lines', which likewise suddenly disappear.

Dip faults—If in crossing a fault plane it is necessary to move against the 'direction of dip of the beds' to find the continuation of an outcrop, the direction of crossing is also to the side of downthrow of the fault. The 'throw' of the fault is equal to (the lateral displacement of corresponding beds) × tan δ, where δ is the angle of their dip (*Figure 5.5*). When the ground slope is in the same direction as the dip of the beds, the difference between the calculated amount of the throw and difference in ground level gives the actual 'throw'; when the ground slope is opposed to the 'dip', the sum of these two quantities gives the actual 'throw of the fault'. Alternatively, from the direction of 'dip' of the beds or from a chronological table, ascertain which of any two beds adjoining one another on either side of the fault plane is the younger; 'then that side on which the younger bed occurs is the downthrow side of the fault'.

Strike Method of calculating the 'Throw'—Example:

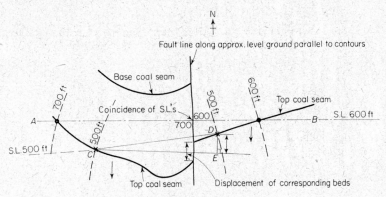

Figure 5.5. Strike method of calculating the 'throw'

160

S.L. 600 ft on west side of fault plane for top coal seam (line *AB*) should meet 600 ft contour on east side of this plane, but with the intervention of the fault meets 700 ft contour here instead. Thus the 'fault' has a throw of 100 ft, *B* being the downthrow side.

Strike faults—hading normal against the direction of dip of beds produce a repetition of their outcrops on either side of the fault line (*see Figure 6.10*). Strike faults hading in the same direction as the dip of the beds cause a disappearance of outcrops due to their submergence, or, if not a complete disappearance, at least a narrowing of the outcrops of those beds which abut against the fault line (*Figure 5.7*).

5.6. EXERCISES

1. Show by suitable sketches how you would identify:
(*a*) a 'pitching fold' on a geological map and in the field;
(*b*) Similarly indicate the appearance, both by plan views and sections, of (*i*) a dip fault; (*ii*) a strike fault.

2. (*a*) By means of suitable map-type diagrams and sections, indicate as clearly and completely as possible the various structures that would be formed in a series of sedimentary rock strata which have undergone regional compression in the past through the process of Dynamic Metamorphism.
(*b*) Describe briefly some characteristic properties of such strata that you would look for in a field survey.

3. (*a*) Explain from a study of relevant maps (e.g., $\frac{1}{25}$, $\frac{1}{10}$ and $\frac{1}{4}$ inch) how Great Britain is divided into various natural geological regions, stating the systems which are represented by each region. Mention any major unconformities and also where systems of 'dykes' and other igneous intrusions occur (e.g., as in southern Scotland).
(*b*) Locate and list the chief rock formations in Great Britain which were formed under non-marine conditions and explain briefly the origins of any 'two' such deposits.

4. Briefly explain the facts a geological map is intended to show; list the sources and scope of information available on the nature and disposition of rocks and soils in British sedimentary terrain, both for 'solid' and 'drift' geology.

5. Describe concisely the geology of an area for which you have personally carried out a field study; mention any exposures of strata or quarries you may have visited and list the various rock specimens you collected for record purposes.

6. A bedded series of limestones dipping due south at 30 degrees outcrop on a horizontal ground surface; the base of the series is visible at the surface 100 yards north of the outcrop of their top bedding plane.
(*a*) Find the true thickness of these beds.

(b) Predict the depth at which the location of the top of these beds could be expected at a position 500 yards to the south of the original upper bedding plane outcrop. What would be your reaction if it was found at a lesser depth than that predicted and what possible explanations would you offer for this?

7. Detail the procedure you would adopt in reporting on the nature and disposition of the 'soils' and 'rocks' in a British sedimentary terrain with (a) practically no drift; (b) an extensive but discontinuous veneer of alluvium, terrace gravel and till (i.e., loose boulder clay).

8. (a) Define the terms 'dip' and 'strike'.

(b) In a series of borings the superface of a bed of shale of constant dip was encountered at depths shown in the following table:

Boring	Coordinates (in feet) South	West	Ground level	Depths to shale (in feet)
A	0	0	208	60
B	600	1,200	197	15
C	1,300	650	203	40
D	700	0	198	?

Find (i) the direction of dip of this bed;
(ii) its angle of dip to nearest $\frac{1}{2}$ degree;
(iii) the depth of the shale in the boring at D.

5.7. PRACTICAL WORK—SOME SUGGESTIONS

From a study of the $\frac{1}{25}$ inch or $\frac{1}{10}$ inch to the mile geological maps of the British Isles and Great Britain, geological sections should be drawn, particularly in the direction from N.W. to S.E. across various parts of the country, in different places from north Scotland down to Wales and similarly from S.W. to S.E. England. Major unconformities should be recognized and extensive systems of dykes, igneous intrusions and lava flows noted. Useful reference can be made to appropriate *Regional Handbooks* for their sketch maps and cross-sections (*see* Chap. 2, Section 2.8).

The distribution and strike of various geological systems will be better understood and the general geological structure of the British Isles interpreted by colouring a series of outline maps separately for each geological system with the standard colour normally used. The various symbols used to indicate different rock types or systems should be listed, and areas composed of specific igneous, sedimentary and metamorphic rocks located for further reference. Note particularly the strike direction of the strata outcrop in the midlands and south-east England.

Diagrammatic outline maps should be attempted to show the relation of topography and drainage to geological structure in regions such as, for example, north and south Wales, the Lake District and parts of Scotland.

Map problems

1. The following borehole and shaft records were obtained from the area shown in the map—*Figure 5.6.*

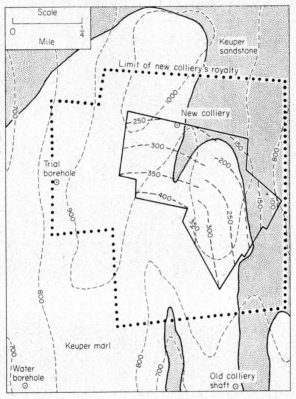

Figure 5.6. Map

New Colliery. O.D. 928 ft

Keuper Marl, to 26 ft deep
Keuper Sandstone, 26–378 ft deep
Coal Measures, 378–668 ft deep
Main Coal, 668–675 ft deep
Coal Measures, proved to 720 ft deep

Trial Borehole. O.D. 866 ft
   Base of Keuper Marl, 150 ft deep
   Base of Keuper Sandstone, 500 ft deep
   Main Coal, 566 ft deep

Water Borehole. O.D. 737 ft
   Keuper Marl, to 232 ft deep
   Keuper Sandstone, 232–583 ft deep
   Coal Measures, proved for 15 ft

Old Colliery Shaft. O.D. 745 ft
   Base of Keuper Marl, 22 ft deep
   Base of Keuper Sandstone, 372 ft deep
   Main Coal, 445 ft deep

The New Colliery has the right to work coal up to the limits shown by the thick dotted line. It has already worked the Main Coal in the area bounded by the thick continuous line, and for this area the strike-lines of the top of the Main Coal are indicated (heavy dashes). The colliery, afraid of flooding from water in the Keuper Sandstone, intends to work the Main Coal only to within 50 ft (vertically) of the Keuper Sandstone Cover.

Indicate on the map:

(*a*) The position of the base of the Keuper Sandstone, by means of strike-lines.

(*b*) The structure of the Main Coal, by means of strike-lines and other symbols.

(*c*) The area in which the coal will probably be extracted within the colliery's royalty.

Illustrate the structure of the whole known succession within the colliery's royalty, by means of sections drawn to scale. (From Mechanical Sciences Tripos Pt. II 1947, Cambridge.)

2. (*a*) Draw a Section along *PQ* on the map (*Figure 5.7*).

(*b*) Describe (*i*) Successions—giving vertical thicknesses of beds *B*, *X* and *Y*.

(*ii*) Structure—giving angle of, and direction of stratum dips and details of the fault.

3. The three formations outcropping in the area shown on map *Figure 4.28* are, in order of superposition top to bottom:

Limestone	*L*
Clay	*C*
Sandstone	*S*

It is proposed to construct a railway running diagonally N.W.–S.E. straight across the area. The track will enter in the N.W. corner at an altitude of 200 ft and rise to the S.E. at a constant gradient of 1 in 60. No cutting should exceed 50 ft in height and the height of each will vary with the nature of the rock.

(a) Mark on the map (or a copy thereof): (i) the proposed course of the railway; (ii) the site of all embankments, tunnels, cuttings, etc.; and by means of an accurate geological section illustrate; (iii) the distribution; (iv) the disposition of the rocks to be encountered.

(b) How far will the excavated materials prove suitable for the embankments?

(c) Give your opinions on the project in a brief report.

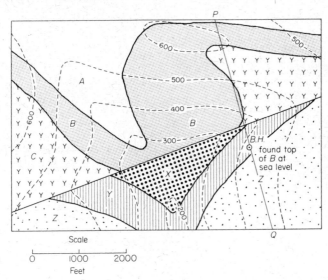

Figure 5.7. Map

4. Give a critical analysis with sections of the geological structure depicted in *Figure 4.29*.

REFERENCES AND BIBLIOGRAPHY

Allum, J. A. E. (1966). *Photogeology and Regional Mapping*. Oxford; Pergamon Press

Bailey, Sir Edward (1952). *Geological Survey of Great Britain*. London; Allen and Unwin

Capper, P. L. and Fisher-Cassie, W. (1963). *The Mechanics of Engineering Soils*. 4th edn. London; Spon

Chalmers, R. M. (1926). *Geological Maps: the Determination of Structural Detail*. London; Oxford University Press

Dowling, J. W. F. and Williams, F. H. P. (1964). 'The Use of Aerial Photographs in Materials Survey and Classification of Land Forms.' *Conf. civ. engng. Problems Overseas. Paper 9*. London: Instn. Civ. Engrs.

Dwerryhouse, A. R. (1911). *Geological Maps: their Interpretation and Use*. London; Arnold

Elles, G. F. (1921). *The Study of Geological Maps.* Cambridge University Press

Flitt, J. S. (1937). *The First 100 years of the Geological Survey of Great Britain.* London; Murby

Fookes, P. G. (1969). 'Geotechnical Mapping of Soils and Sedimentary Rock for Engineering Purposes (Mangla Dam).' *Géotechnique* **19**, No. 1, 52

*Green, P. A. (1968). *Informal Discussion 21232/94,* 12th March 1968—'Ground and Materials Investigations for Road Schemes, Needs and Methods'. London; I.C.E.

Greenly, E. and Williams, H. (1930). *Methods in Geological Surveying.* London; T. Murby

Harding, H. J. B. (1949). 'Site Investigations including Boring and other Methods of Sub-surface Exploration'. *Proc. Instn civ. Engrs* **32**. April

Himus, G. W. and Sweeting, G. S. (1955). *The Elements of Field Geology.* London; University Tutorial Press

Hogentogler, C. A. (1937). *Engineering Properties of Soil.* New York and London; McGraw-Hill

Jenny, H. (1941). *Factors of Soil Formation.* New York; McGraw-Hill

Karol, R. H. (1960). *Soils and Soil Engineering.* London; Prentice-Hall

Keen, B. A. (1936). *The Physical Properties of Soils.* London; Longmans, Green

King, J. H. G. and Cresswell, D. A. (1954). *Soil Mechanics Related to Building.* London; Pitmans

Lowe-Browne, R. (1945). *An Introduction to Soil Mechanics.* London; Pitmans

Nash, J. K. T. Ll. (1951). *The Elements of Soil Mechanics in Theory and Practice.* London; Constable

Nelson, A. and Nelson, K. D. (1967). *Dictionary of Applied Geology, Mining and Civil Engineering.* London; Newnes

Norman, J. W. (1967). *Synopsis of paper on photogeology given at Geological Society of London.* Engineering Group Meeting. Cardiff, 28 Sept. *Proc. Geol. Soc.* No. 1647 (1968)

Penman, A. D. (1964). Editor. *Grouts and Drilling Muds in Engineering Practice.* London; Butterworths

Road Research Laboratory (1952). *Soil Mechanics for Road Engineers.* London; H.M.S.O.

Robinson, G. W. (1936). *Soils, Their Origin, Constitution and Classification.* London; T. Murby

Sharma, R. C. and S. K. (1964). *Principles and Practice of Highway Engineering. Asia*

St. Joseph, J. K. B. (1964). Editor. *The Uses of Air Photography.* London; J. Baker

Whittle, G. and Hepworth, J. V. (1967). *Proc. geol. Soc.* No. 1641, 163

Willox, W. A. (1965). 'Photo-interpretation in Geology and Soils for a Major Road Project in Spain'. *Proc. Geol. Soc.* No. 1629, 5

Williamson, I. A. (1967). *Coal Mining Geology.* Oxford University Press

* *See* relevant references quoted therein.

CHAPTER SIX

ENGINEERING GEOLOGY FOR WATER SUPPLIES

6.1. INTRODUCTION

'Water supply' involves the engineer in the problems of water conservation from superficial sources in reservoirs and the economic use of river, lake or underground supplies.

In this chapter, consideration will be given to the roles of the geologist and water engineer with regard to the conditions required for both underground and surface catchment, and the general meteorological/geological principles which govern such water distribution and collection. The object is to describe the more important geological factors connected with the distribution of underground water in rocks and soils and also to discuss particular conditions which make domestic and industrial water supply a reasonably economic possibility.

6.2. ORIGINS, OCCURRENCES AND MOVEMENT OF GROUND WATER

The origins of 'ground water', which is that occurring within a 'water saturation zone' of variable thickness and depth beneath the earth's surface, and the primary sources of this water, in the geological sense as distinct from the geographical and hydrological viewpoint (e.g., that of ocean waters—into cloud formation by evaporation—then as rain precipitated on to land, flowing into rivers and back to the seas), are threefold:

1. *Meteoric water* consisting of rainfall, dew and snow precipitated on to the land surfaces.

2. *Plutonic water* (also called 'juvenile' or 'magmatic'), usually derived from the condensed steam expelled from molten magmas during their cooling and crystallization, or from lava flows; such water has probably never been in contact with the earth's surface or atmosphere and is generally trapped deep within solidified igneous rock intrusions after their formation below ground.

3. *Connate water*, which is contained in the interstices of certain sedimentary rocks as trapped moisture dating from the time when their rock particles were deposited and consolidated, and their original moisture content was only partially squeezed out of the intergranular voids by overburden pressure.

Of these three, however, the only really significant and principal source of usable underground water for human needs is that listed as No. 1; sources listed under 2 and 3 are, by comparison, insignificant in their total effect on water supplies and compared to rainfall, dew and snow in countries with temperate climates, may likewise be considered as providing a relatively small proportion of the actual *Meteoric water*. In any case, our further discussion of primary water sources will be continued solely in relation to 'rainfall', which, by implication, may be taken to include whatever percentage of dew and/or snow is precipitated on any particular land area.

Occurrences and Movement—Hydrogeology

The permeability of rocks is the main factor in the formation and location of underground water reservoirs, as well as in the movement of water through the *Vadose Zone* into the *capillary fringe* (*see* page 176) and on downwards below the *saturation level* of the water table into the lowest subterranean or '*saturation*' zone. The *water table* surface of separation between the *Vadose* and *saturation* zones is largely a replica of the corresponding surface topography overhead; surface conditions may vary locally from a totally dry state to areas that exhibit swampy or lake sites and lines of springs; the former dry conditions occurring when the W.T. is deep (about 200 ft below ground surface) and the latter wet conditions occurring when the W.T. is regionally shallow, with ground water either near to or appearing at the ground surface. Although the water level in streams and other water courses (termed 'run-off') can be readily noted, such observations may often be misleading as a true indication of the local underground W.T. level, since much subsurface water is absorbed by vegetation and 'transpired', while some of the surface 'run off' water is being continually evaporated into the atmosphere.

Of the rainfall, dew or snow that then falls on any region, the ratio of surface 'run-off' to underground' water retention varies chiefly according to the 'permeability' of the rocks present and the 'nature' of their surface topography, both of which influence the amount of water that reaches the saturation zone. Many aspects of surface water drainage have been dealt with in Chap. 2, and it should be realized that where underground water rises to the surface again as springs, etc., it augments the surface 'run-off', as indicated in *Figure 6.1* and discussed later.

Rainfall precipitated on a land-mass is dispersed partly by 'run-off', partly by 'percolation' into the ground, and partly by 'evaporation' (including its absorption by vegetation followed by 'transpiration'), in approximately the proportions of one-third of the total rainfall to each source of dispersal, when the prevailing climate and vegetation are both of a temperate and moist variety.

The dispersal of rainfall is best represented by the following hydrological diagram (*Figure 6.1*).

6.2. ORIGINS, OCCURRENCES AND MOVEMENT OF GROUND WATER

It is necessary to keep accurate statistics for any locality under consideration as a potential source of water catchment and supply. For this purpose, measurement of the total precipitation in standard rain gauges is used, a gauge being a cylindrical container 5 inches in diameter by 18 in long, partly buried in an open space on level ground, to minimize evaporation, frost and wind effects. The rain is collected in an inner cylinder through a funnel, the top of the gauge being 12 in above ground level in order to avoid the collection of any rain splashes, and the rainfall readings are recorded daily at 9 a.m. G.M.T. for the previous 24-hour period, as indicated in Meteorological Office publications (*see* CP 2001, App. B 300, 400, et seq.) which gives details of national recordings and statistics.

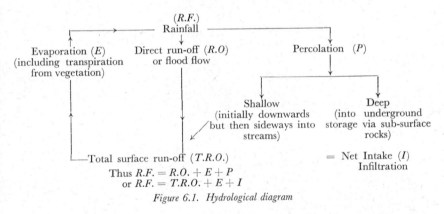

Figure 6.1. Hydrological diagram

From local rainfall records so collected and correlated over a long period of time at numerous judiciously sited stations,* then plotted as inch values, say, like 'spot heights' on a regional plan, 'Isohyetal' lines, joining points of equal rainfall, may be drawn by interpolation to form a Rainfall Map for convenience of summarizing these results pictorially.

Such statistical records and O.S. maps are produced for the British Isles and show that the average annual rainfall for England is 33 in, according to records kept between 1881–1915, and similarly for the British Isles as a whole is 41·5 in. Local variations from these overall average figures, depending on positional and topographic conditions, are very great: e.g., Greenwich has a mean annual rainfall of 24·4 in while Borrowdale in Cumberland has an average of 140 in annually. The variation according to season at Greenwich also shows that, on average, October is the wettest month and February the driest month in any year; monthly rainfall statistics for Greenwich, as averaged over a period of 100 years from 1820–1919, show falls of 1·56–1·59 in during February and March, 1·60–2·0 in from April to June, 2·4 in during

---

* Enquiries to: The Director, Meteorological Office, Air Ministry, London.

July and August, 2·67 in for October and around 2 in throughout mid-winter. Although July and August follow October closely as wet months, evaporation removes most of the higher summer rainfall and in winter it requires 10 in of snow to produce the equivalent effect of 1 in of rain.

For an engineering scheme, local records available must be augmented by the collection of accurate preliminary information as to rainfall and its seasonal variations. An average number of one gauge per 500 acres of catchment area is required.

This collection of preliminary information applies also to evaporation, which is gauged from pans buried and filled with soil to ground level and recordings of percolation are likewise often attempted. The latter quantity is extremely difficult to estimate and percolation gauges* (*see also* Boswell, 1943; Lapworth, 1948; Lewis, 1943; Macdonald, 1961) are not much favoured in practice. There is no way of finding directly the volume of water per unit volume of underground strata *in situ*.

The British Rainfall Organization of the Meteorological Office will calculate total rainfall on a prescribed area for a given period and advise on the siting of gauges. Particular use can be made of the pamphlet *Rules for Rainfall Observers, Form III*.

Again, it is very noticeable from a comparison of the Borrowdale and Greenwich figures how the rainfall varies with land topography, which in turn is based on regional geological structures. The hill formations of the west of Ireland, Scotland and Wales together with the Pennines and the Lake District cause much rain to be precipitated, while the relatively flat area of newer rocks in the east of England have a much lower rainfall. This is, however, partly a consequence of the prevailing westerly winds carrying moist air streams eastward over the British Isles from the Atlantic Ocean.

In considering rainfall on any locality, account must be taken of drought conditions, or periods of exceptionally heavy rain, generally compiled as follows (Table 6.1) for temperate climates:

Table 6.1†
(Statistics gathered over a period of years)

Period	Assumed yield
(a) The driest year	Two-thirds of long-term average annual rainfall
(b) The driest consecutive 2 years	Three-quarters of long-term average rainfall for a 2 year period
(c) The driest consecutive 3 years	Four-fifths of long-term average rainfall for a 3 year period
(d) The wettest year	$1\frac{1}{3}$–$1\frac{1}{2}$ times the long-term average rainfall for a 1 year period

† *See* Glasspoole, 1921–22; 1929–30; 1930, 1947, 1950; 1941; 1949.

* *See* The *Manual of British Water Engineering Practice*, Instn. Water Engineers, 3rd edn, 1961, Chap. 4, page 135, et seq.

This information is required according to local considerations of collection and/or distribution of water on a long-term basis when the statistics for a minimum 3 year period are studied in relation to any water-supply scheme.

The dry weather depletion curves (related to well supplies)* provided for water table maps (*Figure 6.2*), are similarly a great help in this kind of prediction for seasonal variations in ground water levels.

Evaporation—Transpiration. Evaporation of moisture varies seasonally according to the prevailing air temperature and humidity existing over a given region. In Great Britain it is at a maximum in July, with a value as high as 15 per cent of the total annual rainfall, and at a minimum in January, when evaporation of moisture amounts to 2 per cent of the total annual rainfall. The average annual total of this evaporated meteoric water is equivalent to about 16 in of the actual rainfall on Britain, i.e., up to 50 per cent of the total annual rainfall quantity; however, the local variations in this total evaporation as measured, for example, from water level values in open tanks situated at Camden Square, London, may be anything between 10 in and 20 in of rain, according to the relative effects of air and ground temperatures, winds and general humidity in any one year.

H. L. Penman published a map of the British Isles in 1954 (*see* Penman 1948, 1954 and 1956) showing the average annual losses of rainfall (*see also* Lloyd, 1942, 1947 and Nash, 1958) to vary from about 20 inches in southern England and central Wales to about 15 inches in northern England and Scotland. The accuracy of direct measurements of evaporation, whether taken from water surfaces or gauging pans, is low and in the field is complicated by the presence of vegetation, with transpiration and plant usage of water varying widely from place to place. In Great Britain, measurements taken from water surfaces in shallow tanks are probably accurate to $\pm$ 1 in.

Approximately 87 per cent of the summer rainfall total from April to September is evaporated, so that only in the winter period from October to March, are underground water supplies really replenished, when evaporation of precipitated moisture is minimal i.e. 15–20 per cent of the actual rainfall and hence the rainfall available for percolation underground is having its maximum effect.†

Percolation (and hence any Run-off) is particularly affected by the local geology and nature of the ground topography; e.g., steep ground slopes are immediately advantageous for direct 'run-off', while more level ground allows water to lie static, whence it either percolates downwards or dries out by evaporation, according to the precise ground conditions. Vegetational absorption of moisture reduces percolation and plants tend to retain

* D. Halton Thomson has used long-term observations of rainfall to obtain such curves from the water level in wells situated in the chalk of S.E. England and also for estimating percolation quantities. (See for example Halton Thompson, 1921; 1931; 1938 and 1947.)

† *See* Report on Evaporation 1948.

much of the precipitated rain-water; their effect is especially valuable in areas of extremely heavy tropical rainfall as a preventive against surface soil erosion or 'wash-outs'. A clay outcrop or unfissured chalk when once wetted, allows a large amount of 'run-off' or 'evaporation' with very little downward percolation of moisture, while a porous sandy soil produces quite the reverse effect, depending on the amount of water already present in it. Again, any broken, fragmented, or otherwise pervious rock and bare non-porous rock surfaces containing exposed open joints and fissures will generally allow good water percolation, always remembering that 'evaporation' often accounts for the highest amount of rainfall loss. (*See Figure 6.1.*)

Thus, for rocky and impermeable ground, the 'run-off' is above 80 per cent while for sandy porous soil 20–30 per cent is an accepted figure. Forty per cent of the rainfall (Boswell, 1943; Macdonald, 1961) has been suggested as a trial amount for the percolation loss in areas for which no precise data is available. Formulae giving the average amount of percolation over a period of years may also be used and that due to H. L. Penman states, $P = R - 17$ in, where P is the average percolation expected from a measured annual average rainfall of R in. The similar Lapworth formula $P = 0.9R - 13.5$ in for chalk areas only (Lewis, 1943; Lapworth, 1948), is probably the most accurate estimate for a particular type of ground surface but is of limited application; a minimum of 10 in percolation for chalk areas in S.E. England with an average rainfall of over 26 in, is a frequently accepted figure on which to base further calculations.

The student should note that hydrological experience counts for much in making satisfactory estimates of rainfall, run-off, evaporation and percolation over a particular area, as total losses depend on so many hydrological variables and hydrogeological factors from point to point, and the local meteorological conditions from season to season or even day to day. Thus, in practice, for calculating the yield of underground water sources, an estimate is made of the minimum average percolation which could be expected for several years ahead.

6.3. PERVIOUS AND POROUS ROCKS—PERMEABILITY

That any water seeps into the ground at all depends greatly therefore on the nature of surface rock outcrops and whether these are due to 'permeable' types of rock or not. A permeable rock may be either 'pervious' or 'porous' and both terms embrace materials capable of transmitting a water flow.

'Pervious' refers to those types of rock like chalk with joints and fissures which, even in otherwise non-porous igneous or highly-compacted sedimentary and metamorphic types, greatly assist the downward and lateral movement of ground water.

'Porous' applies to rock, such as an open-textured sandstone, with sufficient interstices (pore space) between solid particles to allow the passage or retention of water; this depends on the relative magnitudes of such factors as the particle shapes and their grading, the distribution of any cementing material and the size/shape of pore spaces remaining available for moisture content.

Permeability (or 'the ability of a rock to allow water to pass through it') is measured as the capacity of a rock or soil to transmit water, due only to the effective pull of gravity (or natural pressure head in a local W.T.); relative values of permeability are expressed by the equivalent flow velocities produced in different rocks through unit area and under the local pressure head of a unit (i.e., 100 per cent) hydraulic gradient, as deduced from the measured quantity of water transmitted between given points (Table 6.2).

As shown in Table 6.2, sands and gravels are highly permeable while most clays are practically impermeable*; other impervious rocks include the shales, marls, unjointed igneous rocks, compact limestones and certain residual soil deposits such as the 'Laterites'; these latter often form a blanketing top-soil deposit which prevents the passage of water down to sub-soil and rock beneath. The 'porosity' ratio of a rock is expressed as the

$$\frac{\text{Volume of Voids}}{\text{Total volume}} \times 100 \text{ per cent}$$

a value which varies widely according to the relative influence of the afore-mentioned factors of particle sizes, etc., even in basically similar rock types. As a rough guide to the 'porosity' of rock materials, it should be noted that values > 20 per cent indicate a high porosity, 5–20 per cent (limestone, sandstones) a medium porosity, and those < 5 per cent (granites, quartzites, shale) a low porosity value; values > 40 per cent are rare, except in certain very loose, unconsolidated sandy or other gravelly drift deposits and ordinary open-textured 'top soils'.

Thus, a fairly coarse-grained rock with well sorted and rounded grains, little cemented, holds most water, while a fine-grained rock has most of its pore space filled or cemented. Grain size also determines whether a highly porous rock holds water in its pores by capillary attraction, thus making extraction difficult. 'Wettable micro-porous rocks' like clay (and chalk), though possessing considerable porosity (up to 50 per cent) if dry, become impermeable when saturated (porosity < 59 per cent) because water is held static in their pore spaces by colloidal bond as a film on the clay particles (S.M.R.E., 1952 and Knight, 1952). Therefore, in estimating the quantity of 'free' water (i.e., that available for water supplies), the most important factors and/or properties of a permeable rock are:

(*a*) its 'porosity value' and actual pore sizes;
(*b*) its stratum thickness and the influence of any overlying sediment;
(*c*) its existing degree of saturation;
(*d*) the lowest possible level of downward percolation;

* Typical ranges of permeability values are quoted in Table 7.3.

(*e*) the presence of cracks and fissures in non-porous rock, or 'wettable' micro-porous types.

The Water Table

At great depths however (say 10 miles below ground surface), any joints and fissures are kept closed by the pressure of overlying rock strata; pore space is also non-existent, as the weight of overburden, at such depths of 10–15 miles, crushes the weaker rock grains. There is thus a natural downward limit to percolation and an upper boundary limit to the zone of underground saturation already mentioned on page 168 as the water table. This zonal surface was stated to run approximately parallel to, although generally somewhat flatter than, the topographic surface and its upper boundary level is usually plotted from well measurements at different points, or by such geophysical methods as electrical resistivity survey (Chap. 7), any intervening points required being obtained by interpolation.

Water table maps,* showing the highest and lowest W.T. levels in a region, are produced together with charts to correct for variation of these levels at different seasons of the year. (*Figure 6.2.*) The W.T. surface rises under hills and falls towards streams and rivers, its resulting underground gradients being determined locally by the amount of evaporation and the quantity and rate of water flow through the rocks present, which in turn depends on the effective cross-sectional area of open pores, fissures and joints.

Fluctuation of W.T. level thus depends on the permeability of the waterbearing rock, any rise in level being much less in permeable strata than in those of low permeability. In coarse Bunter Sandstone with open pores, seasonal changes in level summer to winter may be less than 4 ft, but in a slightly fissured chalk, more than 25 ft; although the porosity values of these rocks are similar, here, the pore size, is a dominating factor in water retention, since the chalk is a microporous rock and movement of water through it depends on the influence of joints and fissures.

Rock permeability and water flow—Darcy's law. The relationship between the quantity or velocity of any water flow and the permeability of a given rock is expressed by Darcy's law:

The quantity of liquid passing in unit time,

$$q = A.K_w.i$$

where A is the cross-sectional area of seepage transverse to the water flow in a given rock; K_w is the coefficient of permeability for water, a constant governed by the physical characteristics of the rock material; i, is the hydraulic gradient; expressed as

$$h/l = \frac{\text{Loss of head of water}}{\text{Length of drainage path}}$$

* These maps are available from the I.G.S. and Water Resources Board, Reading, for the chalk aquifers of S.E. England and the Triassic sandstones of the Midlands.

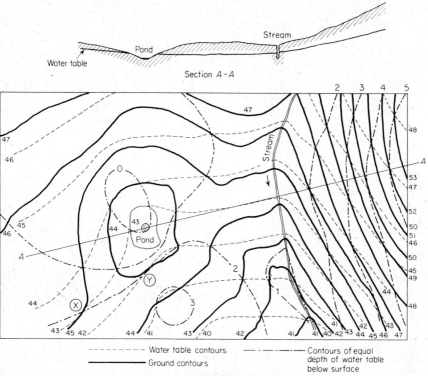

Figure 6.2. Plan showing ground and water-table contours and contours of equal depth of water table below surface

(Note: The results of ground-water investigations are recorded as: (1) Water table shown on the soil profile. (2) Water-table contours over the site. (3) Lines of equal depth to water table below the surface. For ground-water investigations a 25 in to 1 mile or larger-scale plan is desirable. If information is to be given on the depths to the water table, ground contours at 1 ft intervals are required.

(From *Soil Mechanics for Road Engineers*, by permission of the Controller, H.M. Stationery Office)

i.e., $h =$ loss of hydraulic head under gravity measured over distance $l =$ length of water path through a given stratum.

Typical values of K_w for various sediments are given in Table 6.2* and are further discussed in Chap. 7.

Table 6.2. Typical K_w values

Clay	0–0·001	ft/day
⎰ Solid chalk-unfissured	0·042	ft/day
⎱ Silt and fine sand		
Medium sand, sandstone	1·03	ft/day
Coarse sand and fine gravel	25	ft/day
Coarse gravel	100–1,000+	ft/day

* *See also* Casagrande Soil Classification Chart—'Drainage Characteristics' column and App. J 600 and 700 in *CP 2001*. For detailed information on permeability *see* S.M.R.E.

Where the water table intersects the ground surface, water emerges to form either a pond, stream, rivulet, or a general surface dampness or marsh-like seepage over a wider area. In temperate wet climates, the W.T. usually coincides with the topographic surface along the bottom of valleys and directly contributes to any surface 'run-off'. Sometimes seasonal changes of heavy rainfall or drought conditions elevate and depress the W.T. so that intermittent seasonal streams are produced, especially in certain types of rock strata; e.g., the 'Bournes' of the English Chalk Downs, such as the Croydon Bourne which rises at Caterham in Surrey, and Eastbourne, after which this South Coast town is named (*see* Section 6.4 (*c*)).

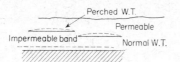

Figure 6.3. Perched water table

A 'Perched' water table (*Figure 6.3*) is produced when a local impervious rock layer holds up a small amount of ground water above the general regional W.T. level. Such an effusion can often mislead the inexperienced water supply investigator by providing a good but only temporary or intermittent supply of water which, if the 'perched' W.T. is shallow, is very liable to serious contamination.

Vadose water—Between the ground surface and the W.T., a zone of downward percolation exists through which 'Vadose' water is periodically, or continually, gravitating inwards to the saturation zone; thus over any particular area, ground water conditions can be very complicated indeed, due to property variations in differing layers of rock (some permeable and others impermeable), or the effects of folding, faults and fissures. Some water may be held static by capillary forces in finely porous rock, forming a zone called the 'capillary fringe' which lies immediately above the W.T. level. It is the combination of features like these which produces the various phenomena we are now to discuss.

Aquifers—First, it will be appropriate to define what is meant by a water-bearing rock stratum of value, one that is given the name, Aquifer (Latin: *Aqua* = water, *fero* = to bear); i.e., any permeable stratum deposited above an impervious stratum and through which water can pass or be retained as a source of water supply from its region of underground storage. A suitable 'aquifer' type rock, when combined with a satisfactory area of outcrop for its water collection from a reasonable topographic catchment of rainfall, can always provide an adequate supply of drinking water.

6.4. SPRINGS AND WELLS

Springs and wells occur as products of the W.T. where ground water flows out naturally at one point on the ground surface, or is available at shallow depths; they are comparable, but a 'seep' which is similar in its effect to a 'bog', is an ooze of water covering a wide area, usually associated with damp, coarse grass as a covering vegetation.

Springs may be classified under the following headings:

(a) A Valley Spring

This spring occurs where the water table rises above the actual ground surface formed by the V-shaped dip of a valley (*Figure 6.4*). Good examples

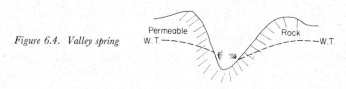

Figure 6.4. Valley spring

of these springs are to be found in the Duddon valley of the English Lake District; after an unusually dry season such springs may also run dry.

(b) A Stratum Spring

This kind of spring sometimes called a 'contact spring', is caused when water passing downward through a permeable deposit is held up by an impervious layer such as occurs often at the contact between the Lower Greensand and Weald Clay in Kent and Surrey. 'Stratum springs' (*Figure 6.5a*), which usually give a discharge dependent upon the severity of a rainy season, also occur as 'overflow' springs when the permeable water-bearing layers dip down beneath an impermeable layer. This happens at

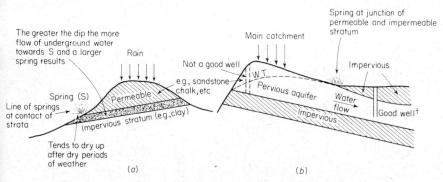

Figure 6.5. (a) Stratum spring (simple type) (b) Dip and scarp topography (overflow spring)

Chalk has micro-pores and although its porosity value is high, this does not mean that water passes through it except by means of cracks and fissures, being otherwise held in the pores. Permeability implies the presence of water channels for a flow through this type of porous rock.
Artesian if water rises to surface; depending on topography. Sub-artesian if water rises in borehole without reaching the surface—depth to water level determined by stratum dip.

Ewell, Surrey, where tertiary clay butts against, and overtops, the chalk strata dipping northwards from their outcrops on the escarpment slope of the North Downs. (*Figure 6.5b.*) Similar occurrences may result at a 'surface of unconformity', when the stratum outcrops behave similarly.

(c) Bournes

Bournes as mentioned previously, are common in chalk areas when the water table rises seasonally after winter rainfall, causing springs and streams to break out at the surface for the earlier part of a year. One such is the Croydon Bourne, which rises at Caterham, Surrey, if sufficient late-season rain falls by midwinter in the last quarter of any year. A 'Bourne' is a special case of the valley-type spring with rather predictable habits regarding its intermittent flow or drought periods.

(d) Fault Springs (Figure 6.6)

Fault springs occur when a permeable stratum stops at a fault plane against an impermeable rock formation; they often form a line along the fault's outcrop and are therefore valuable evidence of the fault direction for mapping purposes. These springs commonly supply a permanent and useful flow of water, but where a fault is filled with an impervious clay 'breccia', the fault channel is sealed, so that not all such cases permit of water drainage or yield a surface supply. Typical 'fault spring' examples are to be seen at Horsted Keynes in Sussex, where there is a fault between Tunbridge Wells and Ashdown Sands, two sandy strata of differing permeabilities, from which the yield is sufficient to provide a continuous public supply.

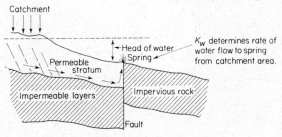

Figure 6.6. Fault spring

(e) Springs From Underground Streams

These are formed along bedding planes with enlarged solution channels and may somewhere come to the surface, expecially so along a valley outcrop, as occurs often in the limestone districts of the Mendips and Pennines.

(*f*) *Artesian Springs*

These springs (named after the province of Artois in northern France) are usually to be found, for example, where Boulder Clay imprisons water under hydraulic head in permeable rocks and the upper impermeable layer is punctured, so that water is forced to the surface and flows out freely (*Figure 6.7*). In general, when a 'synclinal trough' is formed of alternate

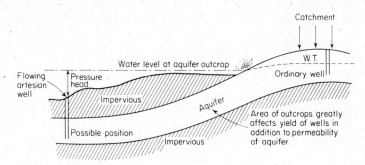

Figure 6.7. Artesian water supply

permeable and impermeable rock strata, with an impermeable cover of newer sediments in the centre, this allows the collection of water under hydraulic head in the basin; then a well sunk in the trough at a lower level than the W.T. at the aquifer outcrops, will give a water flow without pumping. Usually the permeable strata (sandstone, limestone, chalk) is more resistant than the overlying impermeable sediment and the run, or 'scarp', of the trough where the aquifer rock is exposed, is the catchment area; this occurs where the chalk outcrops to the north and south of the London basin.

More examples of these springs/wells will be given later, but as large sources of supply in Great Britain they are generally of infrequent occurrence.

Development and Exploitation of Springs for Public Water Supply

The quantity of water available from spring supplies may be increased by (*a*) digging trenches or inserting pipes to conduct the flow from several points into one channel and thus prevent run-off and percolation wastage into the surrounding earth; (*b*) enlarging the natural outlet of a spring; (*c*) running one or more tunnels or 'adits' into a hillside in order to intersect the W.T. and concentrate the normal underground flow towards one major outlet.

Careful examination of spring water is necessary because of its susceptibility to pollution; the filtered water issuing from a sand and gravel stratum is usually safe for human consumption, but that from limestone or chalk is often contaminated, since it has travelled mostly unfiltered through joints and fissures in these rocks.

Wells

Wells are formed as controlled water supply sources or artificial springs, by digging or boring down to reach water locally in the saturated zone and are considered shallow if less than 100 ft deep. They should be dug below the lowest recorded local W.T. level (*Figure 6.8a*), or their upper shell lined to avoid the risk of surface and vadose water contamination; they will provide a continually adequate, if only small, supply for country houses,

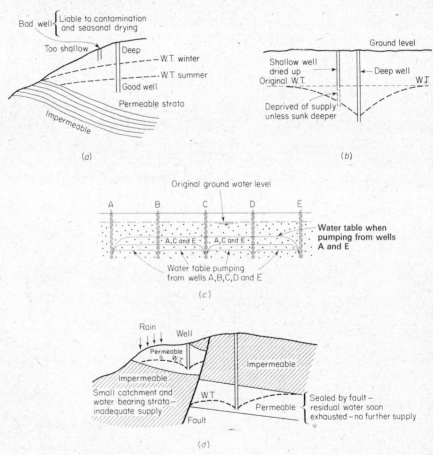

Figure 6.8. (a) *Typical wells.* (b) *Cone of exhaustion.* (c) *Well point system of ground water lowering.* (d) *Unsuitable wells*

((c) from *Handbook of Geotechnical Processes*, by courtesy of Soil Mechanics Ltd.)

farms, etc., so long as care is taken not to exhaust them too quickly or exceptional drought conditions are avoided. As water is pumped continuously, the flow in the aquifer rock may be inadequate to keep pace with an excessive demand and the W.T. in the vicinity of the well becomes depressed, thus forming a surrounding circle of influence which extends downwards in the form of an inverted dry cone known as a *cone of exhaustion* (*Figure 6.8b*). This is the principle used as the basis of 'ground water lowering', a technique adapted to artificially lower the W.T. in the case of an excavation. A deep well may bring a large area into its 'cone of exhaustion' and the overlapping cones (*Figure 6.8c*) in a 'Well Point System'* can dry out waterlogged ground to a considerable depth over the area of a large excavation. Where the water supply in a well is inadequate due to its poor location, the demand must be reduced by alternately resting and pumping at a suitable rate, but the long-term remedy for increased supplies lies in careful siting of wells and deep boring below W.T. level, provided the aquifer rock is satisfactory.

Two examples of unsuitable wells are illustrated in *Figure 6.8d* and follow those indicated also in *Figures 6.5b* and *6.8a and b*.

Determination of K_w Value—In situ Method

In certain cases it is possible to use the following technique in a pumping test of long duration (say 14 days) to determine the value of K_w *in situ*. Water is pumped from a trial well and careful observations of ground water level are made in several boreholes sunk within the vicinity, but at different distances (up to 2 or 3 miles) from the test well; the borehole readings, *when steady*, are used to plot curves showing the lowering of the *water table* (termed *Drawdown*) which corresponds to the recorded rate of water extraction from the well.

The formula

$$Q = \frac{1 \cdot 36 K_w (D^2 - d^2)}{\log_{10}\left(\dfrac{R}{r}\right)}$$

is then applied to find K_w. (Units given on page 182.)

Flow takes place radially into the well and follows the natural bedding planes of the stratum so that the value of K_w is greater than that in the vertical direction.† This indicates that laboratory determinations should be carried out, where possible, on undisturbed samples in directions both parallel and normal to the natural bedding planes; the more appropriate value of K_w

---

* *See A handbook of Site Investigations, Geotechnical Processes and Construction Devices*, Soil Mechanics Ltd., London.

† *Note*: In the ground, the horizontal permeability of a stratum may be much higher than its vertical permeability. Lamination of rock and soil particles in the horizontal direction, due to their original sedimentation, may increase the value of permeability along these planes by as much as 10 or more times the permeability in a vertical direction.

181

could then be used to estimate the probable yield of a well for various assumed values of D and d thus allowing for rock anisotropy.

The radius of influence R of the 'depression cone' may be estimated from experience or found from the borehole observations already outlined. A minimum of two measurements on each of two mutually perpendicular lines set out horizontally are required, usually the lowered water table depth immediately outside the effective well radius r and the normal water table level beyond the radius R. It is, however, preferable to take equilibrium measurements, d_1 and d_2, of the lowered W.T. level in two bores on each line at radii r_1 and r_2 from the pumped well, when the formula:

$$Q = \frac{\pi(d^2{}_1 - d^2{}_2) K_w}{2 \cdot 3026 \log_{10}\left(\dfrac{r_1}{r_2}\right)}$$

would be used (all linear quantities are measured in foot units, and K_w in ft/day to derive the value of Q in ft³/day; or, all linear quantities are measured in centimetre units and K_w in cm/s to derive the value of Q in cm³/s).*

Typical measured values are given as an example of the above approach in Question 7, Section 6.7, to which a solution is given on page 184.

Some experiments by Geological Survey have shown that reliable information on the permeability of fissured strata can be obtained from a series of short pumping tests (Ineson, 1952, 1953; 1959 and 1960). The probable yield of a new well may also be estimated by this method which involves successive increases in the rate of pumping, to give different values of drawdown. In all pumping tests, expert interpretation of results and experienced use of formulae is required. In some problems of ground water flow, increasing use is being made of hydro-geological model techniques, electrical analogues and iterated numerical analysis. (Ineson (1967), and Jones (1968).) Scientists of the former D.S.I.R., working with the R.R.L., B.R.S. and I.G.S., as well as the Water Resources Board and university researchers, are today carrying out many useful experimental investigations.

Theory

Let $Q = $ the quantity of water pumped from the well in unit time (ft³/day) and D ft $= $ the equilibrium depth from the W.T. at radius R, to the impermeable stratum.

Let d ft $= $ the equilibrium depth from the lowered W.T. at the effective well radius r (i.e., immediately outside of well filter) to the impermeable stratum.

Let K_w be the coefficient of permeability of the rock/soil stratum in ft/day.

Consider a vertical elemental cylinder of radius r and thickness dx,

* *See* British Standards 1967 for S.I. units.

below the W.T. level where the depth from the lowered W.T. level to the impermeable stratum is y (*Figure 6.9*).

Applying Darcy's equation, $q = A.K.i$,
when an equilibrium condition giving a steady flow has been established and height of water d in the well is constant.

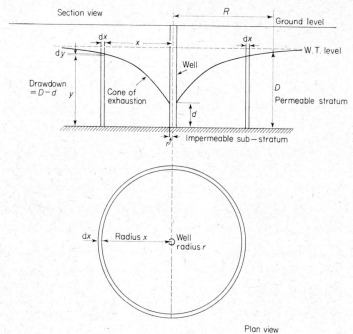

Figure 6.9. Determination of K_w values in situ

Then

$$Q = (2\pi x.y.) \, K_w \frac{\mathrm{d}y}{\mathrm{d}x}$$

[where dy is the loss of head in rock soil path length dx]
and

$$\frac{\mathrm{d}x}{x} Q = 2\pi K_w y \, \mathrm{d}y,$$

Q remaining constant as pumping is maintained.
Hence,

$$Q \int_r^R \frac{\mathrm{d}x}{x} = 2\pi K_w \int_d^D y \, \mathrm{d}y$$

183

by integration,

$$Q \log_e \left(\frac{R}{r}\right) = \pi K_w (D^2 - d^2)$$

Thus

$$Q = \frac{\pi K_w (D^2 - d^2)}{\log_e \left(\dfrac{R}{r}\right)} = \frac{1 \cdot 36 K_w (D^2 - d^2)}{\log_{10} \left(\dfrac{R}{r}\right)}$$

Note: with K_w in cm/s, D and d in cm, Q will be obtained in cm³/s. If Q is required in gallons per minute the above formula becomes:

$$Q = \frac{16 \cdot 75 K_w (D^2 - d^2)}{\log_{10} \left(\dfrac{R}{r}\right)}$$

with all other quantities in feet and minute units.

In practice, R, the outer radius of the cone of depression is often obtained from the formula

$$R = 300(D - d)\sqrt{K_w}$$

where R, D and d are in the English units of feet, but K_w is given in cm/s; or R may be assumed as 1,000 ft for an approximate calculation. The drawdown $(D - d)$ might then be estimated from the fall in water level in the well before and after pumping.

Example (Solution to Q. 7, Section 6.7)

Given,   at $R = 30$ ft, $D = 25 - 3 \cdot 5 = 21 \cdot 5$ ft.
          at $r = 6$ in, $d = 25 - 7 = 18$ ft and $K_w = 3 \times 10^{-3}$ cm/s

then

$$Q = \frac{1 \cdot 36 \times 3 \times 10^{-3}(21 \cdot 5^2 - 18^2)}{2 \cdot 5 \times 12 \times \log_{10}\dfrac{30}{\frac{1}{2}}}$$

$$= \frac{0 \cdot 134 \times 39 \cdot 5 \times 3 \cdot 5 \times 10^{-3}}{1 \cdot 823}$$

$$\therefore Q = 0 \cdot 0104 \text{ ft}^3/\text{s}.$$

6.5. WATER SUPPLIES FROM AQUIFERS

Some of the most interesting and more important British examples are listed as follows in order of age of their geological strata:

(a) *Rocks of the Newer Paleozoic* are good for providing local domestic wells and springs, etc., which give relatively small, but reliable and continuous, supplies of water. For example, Bristol uses the springs from Cheddar by piping water emerging through fissures in the Carboniferous limestone of that region and Coventry takes underground water from wells sunk into the sandstone aquifers situated above the Midland Coal Measures.* Thus the Old Red Sandstone, Carboniferous Limestone, Coal Measure Sandstones (e.g., the Pennant Rocks of south Wales) and Magnesian Limestones all yield useful local supplies, either from springs or wells.

The Carboniferous Limestone, in particular, holds much hard water (due to the presence of bicarbonate of lime) in rock fissures or underground streams, which may be tapped and piped for industrial or town supplies; while the Millstone Grits of the Pennines contain pure 'soft'† water which is best extracted from fractured or fissured rock via springs or wells, to augment the natural surface run-off into the catchment area of many reservoirs. Coal Measure water is mostly contaminated but useful for industrial purposes as large supplies are available in the locality of many industrial towns.

Hard water may be passed through water-softening plants before domestic use to remove the dissolved calcium or magnesium carbonates and sulphates, which are the cause of 'hardness'. Carbonate salts in solution lead to a 'temporary hardness', and produces the typical calcium carbonate ($CaCO_3$), 'fur', lining the inside of a kettle or hot water pipe; but dissolved sulphates produce a 'permanent hardness' in water, which can only be removed by chemical treatment as in a Zoelite water softener.

The foregoing illustrates well the solvent action of acidic ground water on limestone due to the presence of atmospheric carbon dioxide (CO_2) in the original rainwater; the solution process forms calcium bicarbonate with the 'temporary hardness' effect already mentioned and so-called because of the easy manner of its removal from the water by boiling, when the calcium carbonate ($CaCO_3$) is again deposited as a 'fur'.

(b) *Mesozoic rocks* contain some of the largest and most important aquifers in Great Britain. The sandstones of the Triassic (e.g., Bunter Sandstones) supply towns such as Nottingham and Wolverhampton in the Midlands with an ample yield of water, as much as one-third of the outcrop catchment of rainfall being recoverable from these highly permeable rocks, although the diameter of any borehole greatly influences the actual yield obtained.

The Lower Keuper Sandstone overlying the Bunter provides much good water in Warwickshire and Worcestershire, some of it artesian, as at Stratford-on-Avon, where the aquifer rock lies beneath the impervious Keuper Marl (Table 1.3).

Of the Jurassic systems, the limestones of the Great Oolite series in Bedford, Northants, Lincolnshire and Yorkshire provide hard water supplies adequate

---

* Water Supply Memoirs and war-time pamphlets of the D.S.I.R. and H.M. Geological Survey give such local information from 1899–1946 et seq. Information now obtainable from the Water Resources Board, Reading.

† Pure water is automatically 'soft' since it contains no dissolved impurities.

for many sizeable villages and small towns sited on, or near to, any of their outcrops which act as rainwater catchments.

Rocks of the Cretaceous System contain the microporous chalk, the best known and individually the largest of all British aquifers which, although possessing high porosity value, allows water to flow through it only by way of joint planes, cracks and fissures and the water is often very hard. The artesian basin formed by the chalk below London provides many examples of artesian wells; of these the deep wells sunk since 1885 into the chalk under London used to overflow but recently, due to excessive pumping, the water table level in the chalk under central London has fallen progressively at the rate of 3–5 ft each year, having been lowered in all, some 200 ft (*see* Buchan (1938), Skempton (1957) on piezometric levels and also Wilson and Grace (1942)). This chalk aquifer, formed as a sandwich with the impervious London clay above and Gault clay beneath it and folded into the shape of a syncline, is the most typical British artesian example of a shallow synclinal trough structure. Other artesian examples occur in Warwickshire with the Bunter Sandstone of the Midlands underlying Keuper Marl and, in Cambridgeshire, where the highly permeable Lower Greensand and Lower Eocene Sands underly Gault Clay, or in Lincolnshire, where oolitic limestone and Lower Cretaceous Sandstone locally underly impermeable clay beds.

Paris also draws water supplies from an artesian chalk basin similar to that of London; but the most famous example of all is the Great Artesian Basin of Queensland, north Australia, which serves water for cattle stations by means of deep wind-driven pumping wells to a vast area of arid semi-desert covering 600,000 square miles.

Many wells were sunk in London during the last 100 years but the 900 wells known in 1937 were supplying little of the total demand because of the rapid fall in the water-table level following the excessive industrial demand from 1920 onwards. Contamination is likely to occur when the W.T. level in aquifer rocks has sunk to the extent that the flow of Vadose water drains quickly to the wells in an area; this particularly applies to limestone and chalk (in which ground water flows rapidly along enlarged joint planes formed by the water's solvent action) rather than in an aquifer rock in which the underground water flow is slow, thus allowing sufficient time for its purification by filtration. Streams and rivers affected by a lowered W.T. level may lose contaminated water by downward percolation to nearby well supplies if their local W.T. is lowered too far; it then becomes necessary to case wells with a brick or concrete surround, or line them with steel tubes down below the lowest possible W.T. level, so that the water received is thoroughly filtered by its longer passage through the permeable

Note: *Ground Water* (Journal of the National Water Well Association) started with volume 1, No. 1, 1963.

supply rock. The overall drop in the level of the W.T. in the London basin makes the problem of saline contamination from the River Thames and its tributaries through underground fissures into the chalk, a serious one; there is in addition the polluting effect derived from refuse in abandoned or disused wells*.

Further Work of Ground Water

As well as being a contributory geological cause of landslips, a lot of the work of ground water is chemical by nature. Mineral matter is dissolved and so, by interaction, produces chemical changes in other natural substances or minerals; also the mineral compounds carried in solution may be deposited in a variety of places underground, e.g., in mineral veins.

The composition of 'hard' water in limestone regions well illustrates the solvent action of much ground water; in other areas, common salt ($NaCl$), gypsum ($CaSO_4$), various iron compounds and many other chemicals may be dissolved in ground water (e.g., sulphates attack concrete) although their actual quantity is generally very small. Many natural waters often cannot be freed from these 'impurities', except by distillation and when, occasionally, mineral materials are dissolved in considerable amounts, they give the local water a characteristic taste. In some instances a particular type of water is used for medicinal purposes, as at Bath, Harrogate, Leamington, Droitwich and other spa towns in England; and, for example, at Baden-Baden, Karlsbad, Bad Reichenhall, etc., in Germany.

Dissolved iron compounds give water an inky taste and where, in certain spring waters, much iron is present, the rocks over which such water flows become covered with a reddish-yellow coloured coating of iron oxide.

6.6. SURFACE WATER

Supplies from Rivers, Lakes and Reservoirs, Catchment Areas

The major part of public water supplies for most large towns are taken either (*a*), locally from rivers or lakes or (*b*), from water stored in reservoirs on nearby upland areas if possible, and conveyed by pipeline to those towns or more often, industrial cities concerned, each requiring an economic, adequate and nowadays, ever-increasing supply. Some examples of (*a*) in England are as follows:

(1) The Metropolitan Water Board for London now takes 83 per cent (i.e., the majority) of its supply for 7 million people from the River Thames above Teddington,† the remainder being extracted from the natural chalk aquifer underground, or via springs, especially for outlying suburbs south

---

* For a fuller account of London's subsurface water supply see the following references: Boniface, 1959 and 1960; Buchan, 1938, and Skeat, 1961. *See also* Tiffin, 1968.

† *See* Metropolitan Water Board publication: 'A Brief description of the Undertaking with Notes on the Water Works at Ashford Common, Hampton, Stoke Newington and Deptford'.

and north of the Thames Valley, situated on the dip slopes of the Chalk Downlands.

Southampton similarly draws water from the River Itchen and from wells in the local chalk strata.

Darlington also derives its supply mainly from the higher reaches of the River Tees to the west of the town, and these examples are typical of many others.

River supplies, however, require good purification systems, both chemical and bacteriological, but suitable drinking water is otherwise thus cheap and easy to obtain, and may even be taken from a river's lower reaches through water works filtration beds and modern purification plant, which is adequate to deal with any contamination problems.

The Rivers (Prevention of Pollution) Bill, 1960, is, however, intended to prevent any further serious depletion of fresh-water rivers in Britain, just as the Water Act, 1945, allowed control of the abstraction of underground water from over 50 per cent of the aquifers in use. The Water Resources Act, 1963, likewise, has legalized the control of all abstractions of water— whether from surface or underground sources for public or private use— to be permissible only under licence from specified River Authorities acting through a Central Water Resources Board responsible to the Ministry of Housing and Local Government. In future, it seems unlikely that local sources will, as in the past, be developed by a diversity of public and private authorities promoting schemes of their own choice, as listed under (*a*) and (*b*) in this section, but rather that the new River Authorities will supply any natural water to local water boards for them to purify and distribute as necessary*.

(2) One modern British example of natural lake usage is that by Glasgow City, which obtains water from Loch Katrine in the Trossach Hills immediately to the north of the Clyde basin.

Examples of (*b*), involving large water supply schemes in Britain are: Sheffield, with its public water conveyed from the Ladybower, Broomhead and Woodhead Reservoirs in the nearby Pennine/Peak district hills; Manchester's water, carried by aqueduct and pipeline from Thirlmere, Haweswater, Blea Water and Wet Sleddale in the Lake District, and with further provision now being made to flood Langsleddale for increased reservoir capacity; it is now planned to pump water from the natural Ullswater and Windermere Lakes to augment supplies by an extra 45×10^6 gal/day.

---

* A recent proposal has been initiated by the Thames Conservancy Commission in a trial scheme started in 1966 for underground pumping of water from the chalk aquifer above the Upper Greensand, Gault Clay and Corallian at the western end of the Thames valley, the catchment being south of Harwell. It is hoped to augment London's water supply by piping water into the Thames and tributaries from the Vale of Kennet, ultimately at a rate of some 200×10^6 gal/day during dry summer periods (Tiffin, 1968).

Liverpool, which obtains its supply from Lake Vyrnwy and since 1965, also from Capel Celyn Lake (Tryweryn reservoir), these being respectively sited in the Berwyn and Arenig mountains of north Wales.

Birmingham is similarly supplied from the several Elan Valley reservoirs in central Wales, and a consortium of thirteen local authorities from the new Clywedog reservoir, Montgomeryshire.

The largest British reservoirs are impounded by gravity dams (mass concrete or masonry) and were constructed for the Liverpool (Vyrnwy, 1891), Birmingham (Elan Valley Scheme, 1904, with the Claerwen group completed in 1952), and Manchester Corporation (Haweswater, 1941), the Haweswater Dam being unique as the first buttress-type dam built in Britain.

The industrial demand for water is now so heavy for Birmingham, Liverpool and Manchester that distance to the catchment area has become a minor factor in consideration of costs, although the transportation of water from reservoirs so far away requires the costly planning and design of major trunk water lines. Birmingham is perhaps the most fortunate of the three cities named in its proximity to a suitably large catchment area.

On a smaller scale the south Devon coastal town of Torquay and the City of Plymouth both obtain their water requirements from reservoirs situated in the upland granite areas of Dartmoor (e.g., Stithian's Dam Scheme).

Geological, Hydrological and other Factors Related to Reservoir Sites and Catchment Areas

3. In either case (*a*) or (*b*) (page 187) the 'run-off' water is used as opposed to that due to percolation of rainfall and the surface water thus collected, is impounded. This method of catchment requires consideration of various geological factors in choosing a site for a particular reservoir with its attendant dam. For important projects, geologists and engineers are again mutually concerned, the former's exploration and expert advice being essential in uncovering possible future difficulties, which the engineer with his more limited geological knowledge (adequate perhaps for smaller works), may not have foreseen. The cost of an expert specialist site investigation is very small in relation to total costs and the avoidance of a serious error in the siting and design of a reservoir and/or dam (*see also* Section 7.2 onwards). In any case, topographic, meteorological and other factors (such as the statutory permission required, provisions of Water Acts and Orders to be complied with legal cost of way-leave for pipelines, etc.) are all necessary preliminaries to a final choice of site and aqueduct route, for which the engineer may typically need some expert assistance or advice.

4. Some important hydrogeological and other related factors to be considered are as follows:

(*a*) The ratio of run-off to percolation on the catchment area and rainfall loss from evaporation. If they are not available from maps and memoirs,

189

or other references, long-term records must be obtained in the manner previously discussed (Table 6.1).

(*b*) The fact that a reservoir must give the maximum possible water retention consistent with minimum losses by evaporation, percolation, etc.

(*c*) Any dam required must, above all else, be adequate in size, strength, stability and spillway provision.

5. *Hydrological and non-geological factors affecting a catchment area* may be considered thus—a large-scale topographic contoured map will enable the extent of the flooded area to be seen and allow the probable volume of retained water to be gauged or the possible dam positions to be investigated and then comparative excavation, concrete quantities, earthworks, etc., can be estimated.

The choice of reservoir site is also governed by such non-geological factors as distance, altitude, area of catchment, amount of rainfall, etc. (i.e., factors of a mainly topographical and meteorological nature). The rainfall, evaporation, percolation and run-off records needed will be of the type already referred to, with additional information for confirmation of 'meteoric water estimates obtained by local stream gauging, etc.' (Thompson, 1950; Law, 1953. *See also* Lloyd, 1947; Lapworth, 1949; and Nash, 1958.

'Run-off' is measured by river and stream gauges, flumes, notches and weirs and the siting of these is very important. In Great Britain, the local River Boards and the Surface Water Survey Department of the Ministry of Housing and Local Government should be consulted as to the positioning of such measuring devices. The *Surface Water Year Book of Great Britain*, published by H.M.S.O., is also a valuable guide.

The Deacon and Lapworth diagrams (1902 and 1949) are approximations showing reservoir yield/storage relationships for varying annual rainfalls applied to impervious upland catchment areas of England and Wales, and have been much used by water engineers for 60 years since Deacon's original publication. Deacon's diagram and the slightly modified, but easier Lapworth version based directly on run-off, relate to the requisite storage capacity of a reservoir needed to provide a predetermined water-supply yield (Thompson, 1950, and Law, 1953 and 1966; *see also* Lloyd, 1947, Lapworth, 1949, and Nash, 1958) from catchment areas with an average annual rainfall between 30 and 100 in, and assume an annual evaporation loss of 14 in. Corrections for other values are also shown and, in practice, losses of the order of 17 in are often assumed. Both diagrams give a '3 dry-year yield', the usual basis of estimation in British Waterworks practice.

To achieve full exploitation of the catchment area is important and this is usually achieved by providing storage equal to the run-off from the catchment for the driest consecutive 3-year period.

6. Legal matters require careful attention and may even outweigh all other financial and engineering factors of design and construction (Twort,

1963).* For example, consideration must be given to planning and cost aspects of land ownership, fishing rights and amounts of 'Compensation Water' required for riparian owners, wherever involved, in the proposals to the route and cost of aqueduct or pipeline, and also of ancilliary water supplies for other possible users, or those affected by loss or reduction of their previous natural supply to the new catchment area. For such preliminary investigations, the use of 1 in and sometimes 6 in topographical and geological survey maps, together with an organized reconnaissance survey or site visit, will probably suffice for the engineer's personal report on the feasibility of a proposed scheme.

Typical Hydrogeological Data Required

Under headings (3) and (4) (*a*) (*b*) the following details might be noted: 'Over the flooded reservoir area' there must not be any danger of substantial leakage under a full head of retained water, either through drift deposits or the solid rocks.† The positions of both the existing W.T. and possible future W.T. levels should thus be known with reasonable certainty and, in addition, the possible amount of reservoir silting-up must be assessed and estimated, either as being of minor consequence to the design, or the cost of effective preventive measures (silt traps, etc.) for which allowance must be made.

'On the dam site', proving tests by boreholes and geophysical survey for the sub-strata along the line of the dam and a careful inspection of the geological structure are both required so that, wherever possible, the dam foundation may be built throughout in strong water-tight rock and be absolutely safe. The possibility of water percolation creating a hydrostatic uplift beneath the dam and therefore the interrelation of the impounded water level to that of the W.T. around and under the dam also needs to be carefully considered; thus a large scale geological picture (at $\frac{1}{2500}$ say) will usually be required to supply adequate information for these purposes.

Summary

Expert site investigations by specialist firms and geological/soil mechanics site reports are now considered to be an essential feature of all major civil engineering projects and especially so in the case of reservoir sites, each of which requires individual treatment for its own peculiar features. The engineer must have sufficient knowledge to appreciate when to call upon and recognize the significance of the expert advice proffered and know how to act upon it. For example, the whole programming of excavation works will depend on soil and rock conditions and considerable unforeseen problems may arise during construction. Even so, the relatively small cost of

---

* *See* other commentators on the Water Resources Act, 1963.
† Examples are quoted in Chap. 7—'Reservoir and Dam Sites'; *see also* Kisch, 1959.

careful preliminary study—less than 1 per cent of total job costs (e.g., for the Mersey Road Tunnel, 0·2 per cent and for reservoir/dam sites as low as 0·1 per cent up to 1 per cent, depending on the complexity of the works)—is amply repaid by the ultimate savings in design and construction methods. The student water engineer should extend his knowledge by further reading from the references given, but bear in mind the distinctive and individual nature of every reservoir and dam project.

6.7. EXERCISES

1. Discuss briefly the principles which govern the movements of the 'water table' in porous and permeable strata under a land-mass subjected to a temperate climate such as that of Great Britain. Describe some different types of 'springs', with special reference to the geological factors causing them, and outline their merits or demerits in providing a suitable water supply.

2. Describe the requirements of a good source of water supply, relating these in particular to supplies obtained from underground sources such as wells and springs.

Give two examples, with diagrammatic illustrations, of ground water conditions yielding a supply from wells, and two examples similarly for supplies from springs. Indicate in each case the seasonal quantity and quality of supply to be expected from the example chosen with respect to possible exhaustion or contamination and explain why springs are rarely found in certain regions.

3. Give a brief outline of the Mesozoic rocks of England, with special emphasis on the importance of the various formations as sources of underground water. (From I.C.E. Part II examination, April, 1951.)

4. State the geological conditions under which artesian water may be obtained or which are necessary for the formation of an artesian basin and describe in detail the geology of one such regions well known to you. (From I.C.E. Part II examination, October, 1955.)

5. (a) Explain the use of the terms 'porosity' and 'permeability' as applied to water-bearing rocks and discuss in particular the hydrogeological factors controlling the movement of ground water through sedimentary rocks.

Refer to map Figure 6.10

(b) If the bed Y is Sandstone and the other rocks are all impermeable, suggest suitable localities for well boreholes to provide a water supply for the two farms A and B. Give details of the geological sequence likely to be penetrated in each case and draw a section along the line LM.

6. Discuss the terms 'run-off', 'evaporation' and 'percolation' in relation to the dispersal of rainfall and indicate their relative importance in considerations for the siting of a reservoir. What other important geological considerations should be taken into account when conducting the preliminary survey of a possible reservoir site?

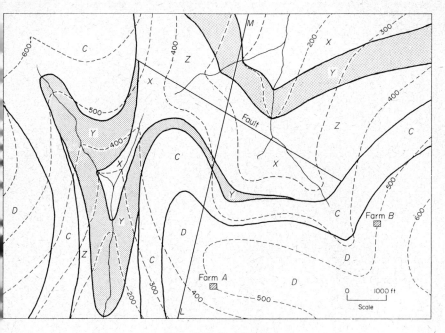

Figure 6.10 Exercise map

7. A well boring with a perforated lining of 12 in diameter is sunk through sandy soil to a level stratum of relatively impermeable soil 25 ft below ground surface. The coefficient of permeability of the sandy soil is 3×10^{-3} cm/s. By steady pumping the water level in the well is lowered to 7 ft below ground level, and at a point 30 ft away from the centre of the well, the ground water level is found to be 3 ft 6 in below the surface. Assuming uniform conditions, calculate the discharge from the well in ft^3/s (*see Figure 6.9*).

If a formula is used, it should be proved, but Darcy's law may be assumed. (From I.C.E. Part II, examination, April, 1956.)

8. A small reservoir for water supply to a local coalfield town is to be constructed by damming the transverse valley from *A–B* as shown on the map (*Figure 6.11*). Typical specimens of the rocks outcropping in the area were collected at the surface for locality 1 (shale) and locality 2 (shale-coal).

193

A trial boring at locality 3 struck a third formation (Sandstone with a gypsum cement) 80 ft down and penetrated 300 ft of it.

By studying the map give:

(a) a general interpretation of the geological structure of the area;

(b) a detailed analysis in the region of the reservoir site. Write a brief report on the merits and demerits of the scheme, excluding at this stage, detailed reference to the dam construction.

(Note: Gypsum is soluble in water!)

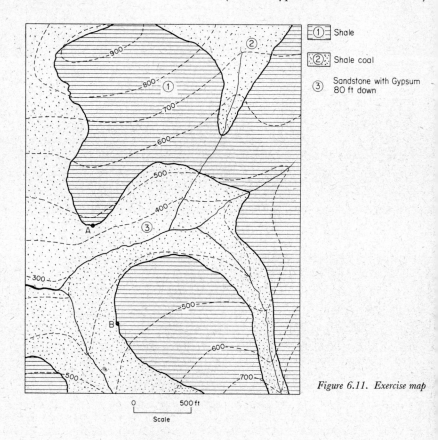

① Shale

②⋮ Shale coal

③ Sandstone with Gypsum 80 ft down

Figure 6.11. Exercise map

0   500 ft

Scale

REFERENCES AND BIBLIOGRAPHY

Anon (1957–8–9). *A Handbook of Site Investigation, Geotechnical Processes and Construction Devices.* London; Soil Mechanics Ltd.

Babbit, H. E., Doland, J. and Gleasby, J. L. (1955). *Water Supply Engineering*, 5th edn. New York; McGraw-Hill

REFERENCES AND BIBLIOGRAPHY

Boniface, E. S. (1959 and 1960). 'Some Experiments in Artificial Recharge in the Lower Lea Valley'. *Proc. Instn civ. Engrs* **14**, 325; **15**

Boswell, P. G. H. (1943). 'The Assessment of Percolation'. *Trans. Instn Wat. Engrs* **48**, 230

Bryan, K. (1928). 'Geology of Reservoirs and Dam Sites'. *Wat. Supply Irrig. Pap., Wash.* 597 (A)

Buchan, S. (1938). 'The Water Supply of the County of London from Underground Sources'. *Mem. geol. Surv.* (Water Supply Memoirs). London; H.M.S.O.

Buchan, S. (1959). 'Artificial Replenishment of Aquifers'. *J. Instn Wat. Engrs* **9**, 111

Buchan, S. and others (1965). 'Hydrogeology and its Part in the Hydrological Cycle—Informal Discussion'. *Proc. Instn civ. Engrs* **31**, 428

Central Advisory Committee (1959). 'Sub-Committee on Information of Water Resources'. London; H.M.S.O.

Comrie, J. (1961). *Civil Engineering Reference Book*. London; Butterworths

Deacon, G. F. (1895–96). 'The Vyrnwy Works for the Water Supply of Liverpool'. *Min. Proc. Instn civ. Engrs* Pt. IV, **126**, 24

Dixey, F. A. (1950). *A Practical Handbook of Water Supply*. 2nd edn. London; T. Murby

Edmunds, F. H. (1941). 'Outlines of Underground Water Supply in England and Wales'. *Trans. Instn Wat. Engrs* **46**, 15

Glasspoole, J. (1921 and 1922). 'The Fluctuations of Annual Rainfall'. *Br. Rainf.* **61**, 228; **62**, 234

Glasspoole, J. (1924). 'Three Driest Consecutive Years'. *Trans. Instn Wat. Engrs* **29**, 83

Glasspoole, J. (1929–30). 'The Areas Covered by Intense and Widespread Falls of Rain'. *Proc. Instn civ. Engrs* **229**, 137

Glasspoole, J. (1930, 1947, 1950). 'The Reliability of Rainfall over the British Isles'. *Trans. Instn Wat. Engrs* **35**, 174; **1**, 441; **5**, 17

Glasspoole, J. (1941). 'Variations in Annual, Seasonal and Monthly Rainfall over the British Isles, 1870–1929'. *Q. Jl. R. met. Soc.* **67**, 5

Glasspoole, J. (1949). 'Seasonal Weather Sequences over England and Wales'. *Met. Mag.Lond.* **78**, 193

Gray, D. A. (1967). 'Hydrogeological Maps'. *Proc. geol. Soc.* No. 1644, 287

Gray, D. A. (1968). Hydrogeological Research at the I.G.S.—B.G.S. Informal Discussion, 11th Dec. London; I.C.E.

Halton Thomson, D. (1921) (1931). 'Hydrological Conditions in the Chalk at Compton, West Sussex'. *Trans. Instn Wat. Engrs* **26**, 228; **36**, 176, and *J. Instn Wat. Engrs* **1** (1947) 39

Halton Thomson, D. (1938). 'A 100 Years' Record of Rainfall and Water Levels in the Chalk at Chilgrove, West Sussex'. *Trans. Instn Wat. Engrs* **43**, 154

Hershfield, D. M. (1962). 'Estimating the Probable Maximum Precipitation'. *Proc. Am. Soc. civ. Engrs* **87**, 475, 99

Hershfield, D. M. (1963). 'Extreme Rainfall Relationships'. *Proc. Am. Soc. civ. Engrs* **88**, 476, 73

Hill, H. P. (1949). 'The Ladybower Reservoir'. *J. Instn Wat. Engrs* **7**, 57

H.M.S.O. (1945). *Water Act.* London

H.M.S.O. (1946–53). *Ministry of Health Summaries of Water Supply Surveys (for County Regions) of England*

H.M.S.O. (1951). *River Pollution Act.* London

H.M.S.O. (1963). *Water Resources Act.* London

Inst civ. Engrs (1950). 'Memorandum on Rainfall—Runoff Relationship and Floods'. *J. Instn civ. Engrs* **34**, 99

Ineson, J. (1952). 'Notes on the Theory and Formulae Associated with Pumping Tests for the Determination of Formation Constants'. *J. Instn Wat. Engrs* **6**, 443

Ineson, J. (1953). 'Some Observations on Pumping Tests Carried out on Chalk Wells' (Symp. on Hydrology). *J. Instn Wat. Engrs* **7**, 215

Ineson, J. (1959) and 1960). 'The Relation between the Yield of a Discharging Well and its Diameter with Particular Reference to a Chalk Well'. *Proc. Instn civ. Engrs* **13**, 299; **15**, 290

Ineson, J. (1967). 'The Role of Resistance–Capacitance Analogue Models in Ground Water Problems'. *Proc. geol. Soc.* No. 1642, 201

Ineson, J. (1968). 'Problems of Ground Water Development'. *Proc. geol. Soc.* No. 1644, 287

Jones, G. P. (1968). 'Hydrogeological Models'. *Proc. geol. Soc.* No. 1644, 290

Kisch, M. (1959). 'Theory of Seepage from Clay—Blanketed Reservoirs'. *Geotechnique* **9**, 9

Knight, B. H. (1952). 'Adsorption of Clays', Chap. 3 in *Soil Mechanics for Civil Engineers*. London; E. Arnold

Lapworth, C. F. (1948). 'Percolation in the Chalk'. *J. Instn Wat. Engrs* **2**, 92

Lapworth, C. F. (1949). 'Reservoir Storage and Yield'. *J. Instn Wat. Engrs* **3**, 269

Law, F. (1953). 'The Estimation of the Reliable Yield of a Catchment by Correlation of Rainfall and Run-off'. *J. Instn Wat. Engrs* **7**, 273

Law, F. (1966). 'Reservoir Yield/Storage Relationship'. *Proc. Instn civ. Engrs* 588. Aug.

Legget, R. F. (1939). *Geology and Engineering*. New York; McGraw-Hill

Lewis, W. V. (1943). 'Some Aspects of Percolation in S.E. England'. *Proc. Geol. Ass.* **54**, 171

Linsley, R. K., Kohler, M. and Paulhus, J. (1959). *Applied Hydrology*. New York; McGraw-Hill

Lloyd, D. (1942). 'Evaporation Loss from Land Areas'. *Trans. Instn Wat. Engrs* **47**, 59

Lloyd, D. (1947). 'The Reliable Yield of an Impounding Reservoir'. *J. Instn Wat. Engrs* **1**, 213

Macdonald, A. T. and Kenyon, W. J. (1961). 'Run-off of Chalk Streams'. *Proc. Instn civ. Engrs* **19**, 23 (Paper 6467)

McDowell, H. R. and Chamberlain, C. F. (1950 and 1952). *Michael and Will on the Law Relating to Water*, 9th edn. London; Butterworths

Meinzer, O. E. (1923). 'Outline of Ground-water Hydrology with Definitions'. *Wat-Supply Irrig. Pap., Wash.* No. 494

Nash, J. E. (1958). 'Determining Run-off from Rainfall'. *Proc. Instn civ. Engrs* **10**, 163

Newberry, J. (1968). 'Influence of Geology on the Design of Dams.' *ICOLD Informal Discussion*, 17th June. London; I.C.E.

Penman, H. L. (1948). 'Natural Evaporation from Open Water, Bare Soil and Grass'. *Proc. Roy. Soc.* **A 193**, 120

Penman, H. L. (1954). 'Evaporation over the British Isles'. *J. Instn Wat. Engrs* **8**, 415 (reprinted from *Q. Jl. R. met. Soc.* **76**, 372)

Penman, H. L. (1956). 'Evaporation: An Introductory Survey'. *Ned. J. Agric. Soc.*, **4**, 9

Road Research Laboratory (1952). *Soil Mechanics for Road Engineers*. London; H.M.S.O.

REFERENCES AND BIBLIOGRAPHY

Report of Research Committee on Standard Methods of Measurement of Evaporation (1948). *J. Instn Wat. Engrs* **2**, 257

Skeat, W. O. and Hobbs, A. T., *et al.* (1961) Editors. *Manual of British Water Engineering Practice,** 2nd edn. London; Institution of Water Engineers

Skempton, A. W. and Henkel, D. J. (1957). 'London Clay in Deep Borings at Paddington, Victoria and the South Bank', *4th Int. Conf. of Soil Mech.* Institution of Civil Engineers

Stow, G. R. S. (1962). 'Modern Water-well Drilling Techniques in Use in the U.K.' *Proc. Instn civ. Engrs* **23** (Paper 6582)

Taylor, G. E. (1941). 'The Haweswater Reservoir'. *J. Instn Wat. Engrs* **5**, 355

Tiffin, C. E. (1968) Editor. 'Success of Thames Valley Boreholes after a Year's Field Trials'. Municipal Journal, 22 March, page 680. (*See also* previous Municipal Journal, 24 February, 1966, page 506, regarding initial scheme)

Thompson, R. W. S. (1950). 'The Application of Statistical Methods in the Determination of the Yield of a Catchment Area from Run-off Data (with a statistical Note by D. Halton Thomson)'. *J. Instn Wat. Engrs* **7**, 273

Twort, A. C. (1963). *A Textbook of Water Supply.* London; E. Arnold

Veal, T. H. P. (1950). *Supply of Water*, 2nd rev. edn. London; Chapman & Hall

Wilson, G. and Grace, H. (1942). 'The Settlement of London due to Underdrainage of the London Clay'. *J. Instn civ. Engrs* **19**, 127

Walters, R. C. S. (1963). *The Nation's Water Supply.* London; Nicholson and Watson

* A standard reference work and particularly good on legal aspects of water engineering.
Note: Water Supply Papers of the U.S. Geological Survey and Water Supply Memoirs of H.M. Geological Survey 1899–1938 give lists of wells etc., and D.S.I.R. Wartime Pamphlets 1940–1946 are useful sources for water supply information.
See also Ministry of Health Summaries of Water Supply Survey 1946 *et seq.*

CHAPTER SEVEN

ENGINEERING GEOLOGY FOR CONSTRUCTION PURPOSES AND DEVELOPMENT OF POWER SOURCES

7.1. INTRODUCTION

In this chapter reference is made to some important applications of geology in the constructional engineering sphere. Let it be remembered, however, that the assimilation of geological data which can be related to engineering problems of construction or to the exploitation of material resources is a necessary prerequisite to any applied study.

First then, a knowledge of *structural geology* is required, with the attendant ability to interpret or reproduce geological maps and estimate the solid shapes, quality and quantity of a given rock material.

Secondly, an adequate knowledge of rock types and their constituent minerals is essential, while 'palaeontology' may also play a significant role in certain instances (e.g., in the context of the *petrology* and stratigraphy of coal seams and oil-bearing strata).

Through such associations *theoretical geology* becomes to the engineer an essential and practical science and in the exploitation of natural resources for modern industry, power supplies, etc., all the diverse facets of *pure geology* may be involved.

7.2. FOUNDATION ENGINEERING—GENERAL

To illustrate the foregoing remarks with particular reference to 'structural geology' the student must by now appreciate that, from an engineering point of view, the geological structure beneath a site is every bit as important as the measured physical characteristics of its superficial soils and rocks.

The foundation engineer today must be especially familiar with the geological aspects of his speciality and the effects of various rock and soil disturbances, natural or constructionwise induced, which he may encounter in practice; for example, many mechanical methods of construction can disturb and dislocate rocks to the extent of destroying their stability, and suitable precautions against possible failures should be considered an essential feature of construction design and practice. In engineering work, lamination, dip or tilt of rock strata, even when slight, and/or the presence of joints or faults and, in particular, shattered, layered and fissured or soft and weathered rocks, may often introduce unexpected complications which would not arise if the strata were horizontal and undisturbed (*see*, for example, Section 6.6—Summary); only small scale features of such a nature, geologically speaking, when unforseen (*Figure 7.1*), can cause endless and costly difficulties during construction, if suitable precautions are not taken in advance. The engineer must also beware of and be able to recognize older geological structures and topographies now concealed beneath more recent superficial deposits. Such structures may not previously have been completely analysed and recorded accurately on existing geological maps, or have been reported in existing survey records of borehole and geophysical observations, or be adequately revealed by such modern methods as *photogeology*.

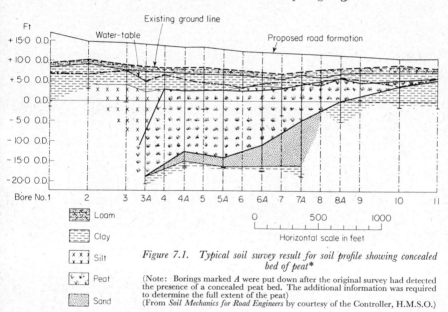

*Figure 7.1. Typical soil survey result for soil profile showing concealed bed of peat**

(Note: Borings marked *A* were put down after the original survey had detected the presence of a concealed peat bed. The additional information was required to determine the full extent of the peat)
(From *Soil Mechanics for Road Engineers* by courtesy of the Controller, H.M.S.O.)

When considering the design of foundation works, a prime factor is the composition and probable future mass behaviour under load of the rock types involved on any possible site; therefore, as the foundation is all-important for a safe engineering construction, intimate knowledge of the structural geology, both 'drift and solid', on a chosen site and an overall appreciation thereof from the engineer's standpoint is vital to the planning of any major works. No hard and fast formulae can be applied, as site conditions may be so variable from point to point that each site must therefore be treated as an individual problem.

* The G.L.C.'s newly opened (1969) Blackwall Tunnel S.A. motorway has an interesting case history in the context of alluvial, peaty ground conditions mentioned on page 46.

Investigations by amply sufficient borings, sampling tubes and trial pits, in conjunction with field classification and loading tests, and laboratory test reports, should indicate beyond reasonable doubt the suitability of the ground, including the effects of ground water, as stable material for any excavations and as a support for the proposed loadings; also, these tests should enable the engineer to decide what construction techniques to adopt and what useful material may be available locally for construction purposes. Clearly, specialist consultant reports will be needed for many sites and no major civil engineering work is now undertaken without its supporting site investigations, the cost of which, as stated in Chap. 6, is marginal in relation to the total costs involved: typical aspects of such investigations are described in Section 7.5, dealing with reservoir and dam sites and in Section 7.8 on Geophysical Surveying. (*See* Green, 1968, and Horslev, 1949.)

7.3. COMPACT SITES FOR BUILDINGS, BRIDGES, DAMS, DOCKS, ETC.

Building Foundations

The free choice of an ideal site in the engineering sense for heavy structures is both rare and fortuitous, as other considerations inevitably apply, such as the available site areas and their evaluation with respect to geographical position, or a structure's estimated economic life compared to its financial return on invested capital; buildings also vary appreciably with each individual client's requirements and these usually predominate, even when or if a choice of several sites is possible. Engineering considerations, where either unconsolidated or consolidated sediments are involved, are described in detail in many excellent books on 'Soil Mechanics' (*see* References, pages 236–237) and in the *C.E. Codes of Practice* (CP 2001–6); here a brief mention only will be made of some important 'solid' and 'drift' geological aspects of various foundation works and the main emphasis will be placed later upon dam construction to provide some examples of the many problems met with in engineering practice.

Great care is needed in boulder clays, for example, not to found piles or rafts for buildings on large boulders giving the appearance of 'solid' rock but still liable to serious settlement; equally dangerous are buried but still under-consolidated clay bands of glacial drift, peat, estuarine silts and river alluvium, etc., within the zone of foundation stress distribution, which again may necessitate the use of piles or a raft-type construction. Fully and over-consolidated strata of gault, lias, London and other clays form good foundation material for rafts, friction or bearing piles when such clays are formed at a suitable depth and free from seasonal changes in moisture content, which may cause alternate swelling and shrinking of these strata; in shallow clay strata dry weather leads to shrinkage and fissuring and wet weather causes swelling by water penetration adsorption. A consequent internal pore water pressure [Skempton and Henkel (1961) and Cooling

(1962)] increase follows with often a dangerous loss of cohesive shear strength (sometimes over a long period of time)* on possible interior slip surfaces (*see* Skempton (1948), Skempton and Henkel (1957), Skempton (1964), Henkel (1957) and Hutchinson (1968)). Igneous, metamorphic rocks like schists and gneisses, well-cemented limestones, sandstones, etc., that is, all hard crystalline or strongly cemented rock types, usually provide strong and

Table 7.1 (*a*). Maximum Safe Bearing Capacities for Horizontal Foundations at 2-ft Depth Below Ground Surface under Vertical Static Loading (From CP 2004, Foundations, page 28. By courtesy of The Institution of Civil Engineers)

		Types of rocks and soils	Maximum safe bearing capacity, $tonf/ft^2$		Remarks
I Rocks	1	Igneous and gneissic rocks in sound condition . . .	100		
	2	Massively-bedded limestones and hard sandstones . .	40		
	3	Schists and slates . . .	30		
	4	Hard shales, mudstones, and soft sandstones . . .	20		
	5	Clay shales	10		
	6	Hard solid chalk . . .	6		
	7	Thinly-bedded limestones and sandstones. . . .	—		To be assessed after inspection.
	8	Heavily-shattered rocks . .			
			Dry	Submerged	
II Non-cohesive soils	9	Compact well-graded sands and gravel-sand mixtures . .	4–6	2–3	Width of foundation (B) not less than 3 ft. 'Dry' means that the ground-water level is at a depth not less than B below the base of the foundation.
	10	Loose well-graded sands and gravel-sand mixtures . .	2–4	1–2	
	11	Compact uniform sands . .	2–4	1–2	
	12	Loose uniform sands . .	1–2	½–1	
III Cohesive soils	13	Very stiff boulder clays and hard clays with a shaly structure	4–6		This group is susceptible to long-term consolidation settlement.
	14	Stiff clays and sandy clays .	2–4		
	15	Firm clays and sandy clays .	1–2		
	16	Soft clays and silts . .	½–1		To be determined after investigation.
	17	Very soft clays and silts . .	½–nil		
IV	18	Peat			To be determined after investigation.
V	19	Made ground . . .			To be determined after investigation.

Note: The values quoted are approximate only and the bearing capacity for individual rocks/soils may differ appreciably from these. It is best to obtain the bearing capacities (particularly of soils) from strength measurements either *in situ* or on undisturbed samples.

* For example, 70–100 years for London clay but notable short-term slip failures occurred at Bradwell, Essex, in 1957. (Hutchinson, 1968.)

Table 7.1 (b). Methods of Test and Sampling for Rocks and Soils
(Means of obtaining samples or access for testing (a) trial pits ⎱ open or timbered
                                        (b) headings ⎰
                                        (c) borings
                                        (d) geophysical survey)
(From *British Standard Codes of Practice*. CP 2001 and reproduced by permission of the British Standards Institution, 2 Park Street, London, W.1, from whom copies of the complete Code of Practice may be obtained)

Type of ground	Methods of estimating	
	Ultimate bearing capacity of ground	Settlement of structures
1. (a) Strong rocks	L	L ⎱ settle-
(b) Weak rocks as shales, weak		L ⎰ ment
limestones and sandstones	F L	L ⎱ small
(c) Non-cohesive soils	F L	F ⎰
(d) Soft compressible soils	L F	L
(e) Stiff, fissured clays	L F	L
2. Soft, compressible stratum overlying hard stratum	L F	L
3. Hard stratum overlying compressible stratum	L F*	L
4. Very variable strata varying in type, thickness and arrangement	Each case to be dealt with on its merits.	

Note: Methods are given in order of preference
F = Field loading test
L = Laboratory tests ⎱ compression and shear tests on undisturbed samples
                       ⎰ consolidation tests on undisturbed samples
                        elastic modulus tests on undisturbed samples.
 * Indicates that tests should be made on each stratum. All tests should be interpreted in light of the geological structure and the surface configuration of the strata. Loading and settlement of similar engineering structures founded nearby on the same strata should be taken into account whenever possible. Reliable information on these in conjunction with tests as above may prove invaluable.

stable foundations. For most structures founded on rock, the only necessity is to prove the depth from ground surface to well into stable bedrock, or explore further into all strata likely to be affected by the foundation loads; for example, to ascertain that there are no soft interbedded soil layers, or in limestones, no solution channels are present, and in coal field areas there are no abandoned unpropped headings existing in the vicinity. Table 7.1 (a) gives typical values of bearing pressures to be used for structural foundations and Table 7.1 (b) indicates test methods of estimating the ultimate bearing capacity of loaded ground and probable settlement of structures.

In 'mining areas' settlement due to collapse of underground workings is a recurring problem, especially where mines are shallow. In salt mining areas also, where rock salt is pumped to the surface in solution as brine, local subsidence usually occurs along the line of any underground solution channels thus formed.

The flow of underground water in certain gravelly type drifts of alluvium

7.3. COMPACT SITES FOR BUILDINGS, BRIDGES, DAMS, DOCKS, ETC.

(e.g., those often vaguely called *hoggin* by engineers) can 'wash out' the finer material, like the silt and sand particles, from such a stratum and cause a similar type of subsidence. A geotechnical process such as that of cement grouting to stabilize and bind these 'movable fines' is often a satisfactory remedy in such cases,* Ischy (1962).

7.4. EXTENDED SITES FOR ROADS, RAILWAYS, CANALS, AND AQUEDUCTS, COAST DEFENCE WORK, ETC.

For planning routes and investigating extended linear sites in order to assess the relative merits of various possible sites and to decide which locations require more careful exploration in relation, for example, to a proposed road and rail alignment, aerial survey methods including 'photogeology' and stereoscopic photogrammetry are most useful. These methods not only allow rapid site reconnaissance and provide a quick means of supplying detailed maps showing the very latest topographic features and man-made obstructions (e.g., new buildings, telephone or electricity transmission lines and pylons, pipe lines, etc.); they also allow interpretation to be made of geological features such as land forms, dendritic and other types of drainage pattern, natural vegetation and sources of constructional materials. Much information has been made available in recent years, especially with reference to motorway construction in England, by specialist firms and surveyors engaged in this class of work,† Green (1968), Whittle and Hepworth (1967), Norman (1967) and Willox (1965).

Cuttings and embankments are usually affected by rock and soil creep or land slips to a degree depending on the size of the works, the condition of the exposed rocks or soil (e.g., their direction of strike and value of dip in relation to the route alignment), and the presence of joints and fissures and quality of drainage. In general, the excavation of all 'solid' rock types, except perhaps chalk, thinly-bedded limestones and shattered or deeply weathered bed-rock, needs the use of explosives and rock blasting techniques (*see* Langfors and Kohlstrom, 1964).

Pervious strata, including many metamorphosed rock types, which rest upon clays or shaly rocks under conditions of dip into a cutting (even if slight in value) and with 'toe' exposure, are of course liable to slip, especially so, if the strata are loaded by earth banks or buildings, roads, railways, etc., which constitute a surcharge.

---

* For details of geotechnical processes used in the stabilizing of St. Paul's Cathedral foundations against underground erosion from the River Fleet and of remedial work at the Bank and Waterloo Underground Stations, also at Bo-Peep Tunnel on the London–Hastings railway line, Branspeth Colliery, Northumberland, and many other cases, *see A Handbook of Site Investigation, Geotechnical Processes and Construction Devices*—Soil Mechanics Ltd. and CP 2004

† *See* Huntings Technical Services Ltd., Brochures (various); e.g., *Geology and Highway Engineering, Photogrammetry* and others including *Hunting Survey Reviews 28* (1962), *29* (1963), *32* (1964). *See also* Mott, P. G. (1963) and CP 2001, No. 1, para. 823, *Site Reconnaissance.*

In the more impermeable soils (Table 7.3), the provision of adequate drainage by a herring-bone pattern land-drainage system, or through rubble-filled conduits, or by vertical sand filter drains (which accelerate consolidation under embankments for example), together with the planting of vegetation to bind the top soil and sub-soil, helps to keep the moisture content relatively constant, to dissipate pore-water pressure more rapidly and hold the G.W.L. low down beneath the sub-surface soil layers; such actions may provide satisfactory remedies against excessive settlement and further slipping, or are certainly preventatives against possible landslides. The effect of changes in moisture content can have vital consequences on the *angle of repose** (or equilibrium slope angle for loose rock/soil material at the ground surface) of slopes; excess water over complete saturation, which can lead easily to a slip, must usually be avoided, especially so in the case of undrained, fissured and weathered clays.

Where a cutting is aligned with the strike of rock strata and is excavated through beds which dip into it, and there is thus a real danger of slip occurring on one side, this side is commonly sloped with the dip of the beds, while the other side may be stable at a much steeper slope. However, for convenience and appearance most cuttings are generally sloped symmetrically with both sides at a safe angle. Vertical walls are seldom used unless the intersected beds are horizontal or dip in the direction of the cutting; tables of safe slope angles for various rocks are available as an approximate guide in most C.E. handbooks and designers' reference publications (e.g., Comrie (1961)).

Some typical values are quoted in Table 7.2.

Table 7.2. Typical Slopes in Cuttings
(From *Soil Mechanics for Road Engineers*, by courtesy of the Controller, H.M.S.O.)

Type of ground	Typical safe slope on excavations
Igneous rocks in sound condition	Almost vertical
Slates, schists, hard shales, hard chalk	$\frac{1}{2}$:1
Thinly bedded limestones and sandstones, mudstones ..	$\frac{3}{4}$:1
Clay-shales, friable sandstones	$1\frac{1}{2}$:1
Gravel and sand	2 :1
Fine sands	$2\frac{1}{2}$:1
Clay and silts	Depends on strength and moisture contents of material and depth of excavation

Note: The slopes given in this table apply to strata with more or less horizontal bedding planes are only approximate typical values. *See also* CP 2003, Earthworks, Table 3 etc.

* The 'angle of repose' is the equilibrium *angle of friction* (or angle at which sliding commences) assumed by a loose granular material at its surface, the precise value of this angle being dependent on the particle sizes, their grading and shapes together with their physical properties of friction and natural state of moisture content.

So generally speaking, for granular materials (such as sands), an earth slope must be always rather less steep than the 'angle of repose' value, if slip is to be prevented and such a non-cohesive slope is then stable to any height; for similar moist materials suitably drained, the repose angle reaches a maximum value and is greater than that for the same soil in a saturated (when the repose angle has a minimum value), or completely dry state, especially since sand grains usually exhibit some cohesion when moist. Cohesive clay slopes do not always, however, exhibit the same repose properties and need very careful consideration, as a sudden rotational slip failure can occur irrespective of their slope angle and height. Even a shallow clay slope may slip (Skempton (1946), Reynolds and Protopadakis (1946)), although in small cuttings some overburden can be removed from the soil slopes and suitable drainage installed, with base retaining walls of minimum height provided as a final support to the banks, if necessary, construction or remedial works being achieved at reasonable cost.

Buttresses and revetments when considered necessary for the construction of bridges and deep cuttings (or when used to reduce the costs of wider excavations, etc., by steepening a cutting's/embankment's side slopes), can be made satisfactory where adequate drainage is provided; this is usually done by constructing weep holes, to pass through such walls at their base, as outlets from a rubble backfill drainage system or, alternatively, by providing drainage tunnels leading underneath their bases from the wall backfill. It also should be clearly understood that although a structure may be a safe design within itself for overall stability down to foundation level, in clay soils particularly, a foundation failure (usually caused through a critical, but often a progressive long-term deterioration in a soil's local strength state by softening via its internal moisture content and pore-water pressure build-up can still occur by deep rotational shear slip on a roughly cylindrical surface extending far below the actual foundation (*see* Skempton (1948), Skempton and Henkel (1957), Skempton (1964), Henkel (1957) and Hutchinson (1968, 1969)).

Landslips (1) and Landslides (2) (see Skempton, 1964; Bishop, 1969; Hutchinson, 1969

Enough has been stated already to merit a few special comments on the mechanism of 'slips/slides', and to summarize the previous remarks, before proceeding further.

1. Sub-surface water is a major agent causing a sudden mass movement of loose ground or rock and soil creep, by increasing the pore-water pressure in between particles to a critically dangerous value, when the internal shear strength may be neutralized, and by lubricating particel surfaces; water thus reduces both cohesion and friction between adjacent layers of soil, and adhesion at the interface of strata. A wet soil mass then creeps or slips,

often as a mud run downhill under the action of gravity.* An undrained sloping clay layer is especially dangerous if thoroughly wetted by percolating rain-water (or saturated by spring water†) from above, which soaking can lead to the large rotational slip failure of the whole soil mass noted in the preceding section. (*See* Soil Mechanics' reference books listed on page 236, for detailed discussion of this deeply based type of circular slip mechanism, *see* CP 2003. *Figure 1.*)

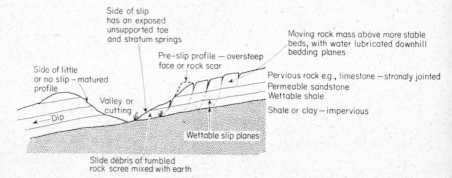

Figure 7.2. Typical landslide conditions

(Section of valley to illustrate slip of jointed rock mass—fractures opening to allow exposed rock mass to break off)

2. With jointed solid rocks (e.g., limestones, chalk, siltstones and flag-stones) the outer surface parts forming a valley or cliff face tend to break away from the main rock mass; this type of landslide is common in the valleys of South Wales,* where hill crests are formed of sandstone which rests on interface strata of shaly-clays or coaly-shale at their base (*Figure 7.2*). As mentioned in Chap. 2, coastal landslides or cliff falls are fairly common where the upper rock strata of a series are permeable, the lower rock strata impermeable, and all dip seaward; on the south coasts of England and the Isle of Wight, chalk cliffs frequently slip in this way, especially when a water-lubricated interface of gault or lias clay underlies the dipping chalk strata.‡

* The South Wales village of Graig-cefn-parc was abandoned in December 1965 after heavy rainfall had caused the creeping hillside on which it was built to slide dangerously and extensively.

† Exemplified by the huge and disastrous fluid mud and flow slide of a shaly coal-waste tip resting on fissured sandstone and clay at Aberfan, S. Wales, on 21st October 1966, and caused, after heavy rain, by an increase of P.W.P. in the presence of spring water at the base of the tip to a critical value which led to its bursting and sudden mass failure (Bishop, 1969).

‡ *See Engng News Rec.* 75 (1916)
    Panama Canal Landslides and other cases.
    I.G.S. Geological Museum Slide A 9673 of Landslip, The Warren, Folkestone, Kent and Hutchinson (1969).

Tunnels

For tunnelling projects, the engineer requires especially reliable and accurate information on the exact geological strata and ground conditions, for estimating the tunnel loads and rock pressures in order to decide the best and safest methods of carrying out such work below ground level. (This applies also to a lesser degree with much special and heavy foundation work, e.g., in the use of 'under-reamed piles'.) Detailed knowledge of the rock stratification and anticipated ground-water conditions to be met should indicate whether the rock material will be difficult to excavate and hence costly to remove (e.g., with much unwanted 'overbreak'—see the following paragraphs); whether the entrance excavations, tunnel portals and headings will be stable if unsupported or otherwise require much shoring and perhaps some special drainage provisions; whether and where a tunnel lining may be necessary or special geotechnical processes, such as ground water lowering, cementation, etc., will be of value. (*See* Toms, 1953, and Széchy, 1966.)

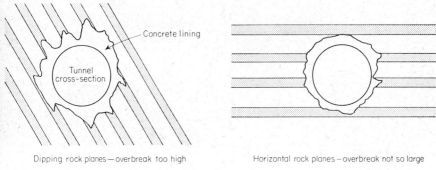

Dipping rock planes—overbreak too high   Horizontal rock planes—overbreak not so large

Figure 7.3. Effect of overbreak in tunnelling

From exploratory observations (geophysical or otherwise) on bedding, jointing, cleavage and hardness properties of the rocks to be encountered, estimates can be made of any possible major difficulties liable to arise during the actual construction stages. In igneous rocks, for example, any 'over-break' due to the use of explosives for rock blasting can usually be kept to a minimum and these rock types, which present relatively little difficulty will stand unsupported in both cuttings and tunnels; but for hard meta-morphic rocks (already largely disturbed or sheared, and folded or faulted during their formation as well as possessing perhaps a 'schistose' character) in which the engineer must resort to 'blasting' to ensure tunnelling progress through them, 'overbreak' (*Figure 7.3*) is often excessive and the more so as the line of driving is orientated with the 'strike' of these rocks.

In the Lochaber Power Scheme, Scotland, 'overbreak' was as high as 40 per cent of the design cross-section in the water-power pressure tunnel when the mica-schist planes were nearly vertical, but only 16 per cent when these planes were horizontal; this tunnel was later lined with concrete to give a final inside working diameter of 15 ft. (*See* Naylor, 1937.)

Younger strata often present problems of slip and flooding which are worst in sands, clays and glacial drift, while generally speaking older massive bedded rocks favour drilling and tunnelling operations and difficulties only increase with water percolation and complicated geological structures or certain adverse rock properties. Thus, with metamorphic types or shaly rocks having a dip downwards into the tunnel, an unsupported roof can lead to severe slips along water-lubricated planes; special drainage provisions and pumping facilities to take the influx of ground water are often necessary in limestones and sandstones, interbedded hard and soft pervious strata of variable texture, and when faults are met.

References to some of the latest tunnel projects are given in Section 7.10— Further Work. (*See also* Prosser, J. R. and St. C. Grant, P. A., 1968, on the Tyne Tunnel, and Kell, J., 1963, on the Dartford Tunnel.)

7.5. RESERVOIR AND GRAVITY-TYPE DAM SITES

To amplify the remarks of Section 6.6 in more detail, we will commence this section by first considering some features of reservoir sites, commencing with the investigation of drift deposits.

Drift Deposits

If, within a reservoir's flooded area, there are deposits of peat, alluvium or glacial clay and till, etc., on top of the solid rocks, these deposits need a very careful inspection through an extensive borehole programme via the study and use of 1 inch and 6 inch drift and solid maps; seismic or electrical geophysical survey methods are now available to ascertain the topography of the solid rock boundary surface beneath the drift and for producing a contoured map of this solid, as exactly as may be felt necessary. Peat must usually be removed almost completely (e.g., as for the Alwen Reservoir, N. Wales, in 1917), because of its doubtful depth when thick and the water pollution arising out of its acidity and organic content, undesirable impurities leading, for example, to coloration of the impounded water and to attack on concrete and water pipes.

In the Silent Valley reservoir, Belfast (*see* Chapman (1923), Halcrow (1928) and McIldowie (1934)), 70 acres of peat present was successfully and cheaply covered over with a 2 ft thick blanket of sand from nearby deposits to give a satisfactory and non-polluting reservoir bed. A sandy and silty alluvium sometimes gives trouble during construction on a dam site because of its variable properties, high-water content and water-bearing pockets, thus affecting the methods of timbering, sheetpiling or otherwise supporting excavations (e.g., by using a Bentonite fill), especially when 'running sand' is encountered. Glacial deposits (e.g., boulder clay) can be useful when impervious, but morainic material (and especially open-textured 'tills') containing sand and gravel may develop a localized leakage with the dangerous possibility of 'piping'.* This particularly applies to lateral moraines which are generally more gravelly and more porous than base moraines.

Again it must be noted whether 'solid' rock discovered in a borehole is not a huge boulder set in glacial drift or a form of hard, consolidated drift cemented by iron compounds, such as 'limonite' which may be soluble under the head of impounded reservoir water.

Solid Rock Structure

Detailed 1 inch and 6 inch maps are necessary in conjunction with careful site investigations planned to ascertain the suitability of the solid rock structure. A simple geological structure is most advantageous; highly folded and complex formations with joints or bedding plane weaknesses allowing possible sliding or percolation are undesirable, especially on a dam site. Also a very useful economy in the cost of dam building can be made if glaciation has caused rock bars by gouging the floor of a valley (*Figure 7.4*).

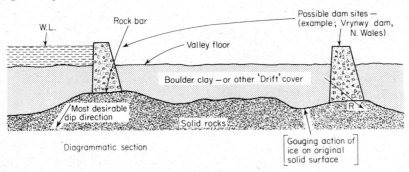

Figure 7.4. Glaciated rock bars

(Clays cover rock bar but after excavation of the drift, the solid bar could be used as a dam foundation at considerable saving in work and costs)

Generally, a dip upstream (to oppose sliding and downhill percolation leakage) or one across the valley, is more desirable than one downstream, and any porous rock beds outcropping on valley sides below the level which will be flooded, need investigation, especially if they dip outwards and downwards through the surrounding hills; a porous bed is often sealed by

---

* 'Piping' can be controlled by constructing a special filter drain of coarse-grained material based on a carefully graded particle size distribution and the possible failure of trenches in fine sand stratum, or the collapse of a dam by under-seepage thus prevented.

cutting a channel over the zone of its outcrop and filling this to a considerable depth with solid dense concrete or by blanket cement grouting of the rock itself, as at the Dokan Gorge Reservoir, Iraq. (*See* Binnie, 1959.)

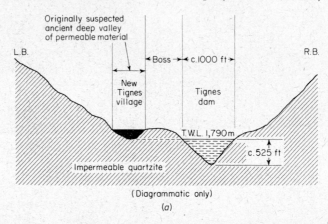

(Diagrammatic only)

(*a*)

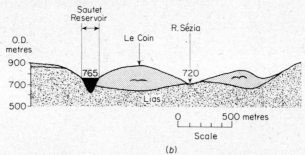

(*b*)

Figure 7.5. Buried valleys near reservoirs in Southern France: (*a*) *Tignes Dam. The level of the bottom of a pre-glacial valley determined the top water level of the dam.* (*b*) *Sautet Reservoir. Considerable leakages occurred from this reservoir through an old valley filled with glacial sands*

(From *Muck Shifter* by courtesy of the Editor)

With regard to the 'solid rocks', those such as shales, slates, schists, gneisses and granite may generally be considered watertight; but again, a highly fissured permeable rock will reduce water-retention although this is always related to the local W.T. level: limestones and similar soluble rocks are poor prospects for this aspect of reservoir design and beds or pockets, layers and cements of gypsum are very soluble indeed. The Hales Bar Dam site on the Tennessee River, U.S.A., suffered badly from the leakage defects of limestone solubility and cavitation and the Broomhead Reservoir, Sheffield, England, suffered from underground leakage through a band of porous grit. Leakage through fissures may also become a serious problem, although these sometimes seal themselves by the silting-up of the bed in many reservoirs. (*Figure 7.6.*)

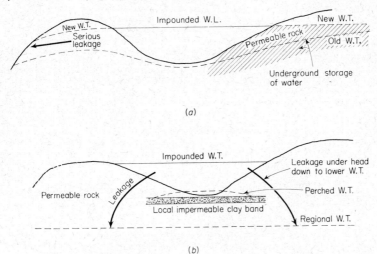

(a)

(b)

Figure 7.6. Leakage of water from reservoirs: (a) Cedar River Reservoir, U.S.A. (see U.S. Wat. Supply Irrig. Pap. Wash. No. 597A, 1926). (b) No. 3 Reservoir, Carlsbad, U.S.A. (see U.S. Wat. Supply Irrig. Pap. Wash. No. 580, 1926.)

Pervious faults are often a major problem, either (*a*) allowing leakage of water from the reservoir, or (*b*) causing the formation of springs during construction of a dam. In case (*a*), wells or sumps must be sunk at the springs and the ground kept reasonably dry by continuous pumping, or by providing temporary drainage channels during the construction period, especially in excavations.

Unstable ground liable to landslips should be avoided,* for example, argillaceous rock types, which when wetted, swell and lose cohesive shear strength, and metamorphic rocks with slate-like cleavage or mica-schists with foliation planes dipping into a valley. In known earthquake regions of the world, special construction techniques are adopted to ensure the safety of dam structures if water must be impounded by such artificial barriers.

Thus, highly porous sandstones and conglomerates and soluble rocks like limestones, or those with gypsum and iron cements and/or veins must be avoided; similarly, certain basic igneous rocks are often not as strong as they at first appear because their iron compounds may oxidize and hydrate with water retention and this leads to disintegration of the rock, e.g., through decomposition of iron pyrites to limonite and hydration of iron magnesium

---

* This caused the Vaiont Reservoir and Dam failure in N. Italy, 1963, *see Proc. Instn civ. Engrs* **28**, July (1964) and *The Guardian*, Oct. 12th, 1963.

silicates. Hard crystalline rocks of the acid igneous plutonic type such as granite, or schists and gneisses (after taking care to investigate their cleavage planes for alteration or slip) and slightly metamorphosed sediments such as shales and slates, are usually suitable for both reservoir and dam site areas.

The Water Table and G.W.L. Observations

If the W.T. is naturally lower than the impounded water, loss by seepage may result, particularly if the rocks are an open-textured variety or extensively jointed. The existing local W.T. level should thus be at approximately the proposed level of the future impounded water on the sides of a reservoir valley, but any perched W.T.'s must be detected and clearly identified as such (*Figure 7.6*). It should also be remembered that the natural dynamics of the reservoir region will be altered by the new water levels and regrading of catchment rivers, etc., so that the equilibrium of the geological rock structure may later be disturbed by changes in the environmental condition of rock masses, such as the effect of pore and joint water pressures, to the extent of serious landslides occurring. Disequilibrium is also a special problem initially; during excavations in sedimentary strata with alternate layers of porous and clayey rocks and in foliated metamorphic rock types (e.g., schists), where separation and slip occur along their natural cleavage planes. An example of dynamic change causing leakage is shown in *Figure 7.6(b)*.

*Ground water level observations**—As indicated in Chap. 6 for the *in situ* determination of permeability of a stratum, observation boreholes or tube wells are the easiest method of obtaining information on the position of the water table (*see* Harding (1949), Cooling (1962) and Skempton and Henkel (1961)). The procedure is to place, at the required stratum level, a tube about 2 inches in diameter and perforated at its bottom, in a borehole 4–6 inches in diameter and pack fine gravel as a filter around it at this level within the stratum to be observed. The space between tube and sides of the borehole, directly above the tube perforations and in the permeable stratum under test, is sealed with puddled clay or bentonite and the top few feet of borehole are also sealed to prevent the ingress of surface water down the bore. Usually the tube projects above ground level and the water level in it is determined by means of a plumb-line sounding or electrical dipmeter.

Observations must be continued long enough for equilibrium levels to be established, which in silts and clays may take some weeks, so that quick measurements made during the borehole sampling of a stratum can be misleading (*see* Penman, 1961).

Methods other than the *in situ* well test are used for determining permeability, such as the laboratory determination of K_w on samples or by introducing electrolytes or dyes into a well and observing the time taken for

* *See* CP 2001, J 600 and J 700.

the flow to reach other wells downstream (see Muskat, 1937); but these are only applicable for noting approximate values and are unsuitable when the flow velocity is low. For some observations of the water pressure in permeable strata and where these measurements require a quicker method of recording changes than is possible by tube wells (for example, changes in water pressure in peats and clays caused by excavations or superimposed loading of embankments etc.), pore-water pressure cells (piezometers) are used. (See Cooling (1962), Skimpton and Henkel (1961), and also Penman (1956; 1961).)

The high pore-water pressures often developed during the construction of earth dams and embankments, for example, may endanger their stability. Today it is a common field practice to measure these pressures in every type of earth dam when under construction and subsequently, at intervals over a long period of years after completion of the work, to check by piezometric observations that the factor of safety is adequate and to obtain data for the improvement of future designs. As often with gravity dams, an impervious cut off is an essential feature of earth dam construction. One simple procedure is to place a porous piezometer tip 2 inches in diameter and 10 inches long, with a connecting $\frac{1}{2}$ inch diameter P.V.C. access tube rising to above G.L. in the bottom of a borehole of 4–6 inches diameter and pack a fine gravel filter around the tip. The space above the tip with its central access tube is sealed as outlined previously for a 2 inch tube and dipmeter water level readings are then taken similarly in the $\frac{1}{2}$ inch P.V.C. tube.

Table 7.3. Ranges of Permeability Value for Soils

Coefficient K_w		Rock or soil type	Remarks
cm/s	ft/day		
1–100	2,360–23,600	Gravels and conglomerates	Good drainage
10^{-3}– 1	2·36–2,360	Sands and sandstones	Includes fissured or weathered clays
10^{-6}–10^{-3}	$2·36 \times 10^{-4}$–2·4	Silts and siltstones	Poor drainage
Zero to 10^{-6}	Zero to $2·36 \times 10^{-4}$	Clays and shales	Impervious

Permeability formulae are derived from the basic assumption that a permeable stratum is homogeneous, while in fact most are variable in character. Ranges of permeability (Table 7.3) of a given material may be large and for a range of materials, very large; alluvium is particularly suspect in this respect. Thus the flow system adopted for testing must be carefully related to that expected in the final construction, otherwise the results obtained may be invalid; erratic drawdown in wells and inspection of borehole samples will help to indicate the presence of heterogeneous deposits.

213

Silting Up

Consideration must be given to the possibility of sediment carried by the streams of the watershed silting up a reservoir with a consequent loss of ultimate storage capacity (*see* Winter, 1950–51). Provision must be made also to allow the sediment to pass through the dam and for placing silt traps on the inflowing rivers to slow down the rate of sediment deposition on the reservoir bed.

Dam Types

The type of dam most suitable for a particular site is largely dependent on the nature of the foundation rock. Two main dam types may be considered: (*a*) *Gravity dams* (of concrete or masonry), (*b*) *Earth and rock fill dams*, while concrete arch and prestressed concrete dams are further special possibilities (*see* Skeat and Hobbs, 1961, Chap. 6, page 230 et seq.). Such remarks as now follow, refer mainly to the construction of traditional gravity dams, since earth dams, although many in number, come more into the province of Soil Mechanics and British practice today deals largely with the construction of gravity or buttress sections for concrete dams. In Britain, future trends are towards the design of more economical buttress-type and combined gravity-arch or arch-buttress type concrete dams, while progress in the pre-stressed concrete field is developing rapidly, especially for hydro-electric dam schemes in Scotland.

Dam Sites—Investigation and Planning

As mentioned previously in Chap. 6, a large scale (say 1/2500) geological map of the dam foundation area will be needed and, as planned from this map, 2 in or 4 in diameter trial boreholes, and later, large bores should be sunk to reach well down into impermeable strata underground and thus reveal the essentials of the overall rock structure. Large boreholes of 3 ft diameter allow visual inspection of the various rocks at depth, as it is vital to know the true nature of the solid rocks completely (especially if fissured and containing a weathered zone beyond and below the dam's foundation level*) and sufficiently far down into sound bedrock; any faults, slips, fissures, etc., and any loss of water level in the boreholes should also be carefully noted and suitable leakage tests made. Geophysical methods may be used to supply information on the rock strata between boreholes and where the strata are, for example, variable, important discontinuities such as pockets of quicksand should thus be revealed.

The dam must be strongly founded in sound bedrock, which should be reached after all superficial deposits have been removed along its line and over its full base width, and blanket grouting may sometimes be used to

---

* A particularly interesting case involving different foundation conditions in recent years is the Wet Sleddale Dam, *see* Walters, 1964, p. 46.

strengthen weak rock. The methods of retaining any lateral superficial deposits adjacent to the foundation trench during construction and dealing with the water in the dam trench excavations will need careful planning, and hence a proper soils classification is required for these deposits. For example, buried 'running sand' and silt were not discovered by the trial borings taken on the dam site (in this case for an earth dam) at the Belfast reservoir mentioned previously, page 208. These faults involved the engineers in costly below G.W.L. constructional methods to prevent anticipated earth slips along with 'piping' and heave of the dam and cut-off trench bottom. Geophysical methods of site investigation would probably have indicated the presence of the troublesome deposits and another site for the dam could have been chosen. In the event, the work was successfully completed, after de-watering the site and the use of compressed air within cast-iron linings for work proceeding in the cut-off trench. But excavation in this trench 212 ft deep) and the final depth of grout curtain required (275 ft to sound bedrock) far exceeded all expectation (and economic financial estimates) and this proved to be the most difficult construction yet encountered in British experience.

If glacial drift covers the site, boreholes must be carefully chosen to give the fullest and most reliable information needed, here also in conjunction with geophysical surveys, as the presence of solids in such drifts can be so very misleading. To quote again a classic example, large boulders in boulder clay or glacial tillite can be too easily mistaken for the solid rock bed and thus geophysical survey methods may again be advantageously used to map the topography of the buried solid surface.

Foundation of dam—The age and state of the rock strata and types must be clearly determined, including their local structural formations and variations of dip; natural moisture content, strength in crushing and shear, and the presence of folds and faults, joints and spacing of bedding planes—all these features need careful observation, tabulation and recording. For the larger gravity dams, granite, strong hard sandstones or gneissic rocks are necessary to carry the foundation loads; for all dams complete records for the compressibility (both wet and dry), natural moisture content, degree of consolidation, and anticipated properties in a saturation state of the solid rocks right along the dam line, as ascertained by rock/soil mechanics laboratory tests, are required.

As with the valley sides of a reservoir, rocks with alternate hard and soft layers are suspect; the layers may too readily lead to slipping and this weakness of laminated rocks like schists, which are otherwise very strong, must be borne in mind in regard to a dam foundation.* Fracture zones of

* Failure of the Austin Dam, Texas, was attributed to water percolation causing slip of clays and shales underlying limestones which allowed solution channels and springs to form.
The Vaiont disaster of 1963 in N. Italy was caused by enormous landslip, *see* References pages 211, 231 and 236.

intense criss-crossed faulting and strong joint systems may be lines of weakness and percolation channels for water circulation; and sharp short wavelength local folding may provide weak points of shattered metamorphosed rock, which should if possible be avoided, but single large faults may not necessarily be dangerous. Recent faults, however, must be avoided although if an existing relative movement should be found along the chosen dam line (no known slide or throw having occurred there during an estimated time of the last half to one million years at least) it can be allowed for in the design of a dam. However, percolation through any fault plane breccia must be investigated, even if, as often happens, the plane is silted up.

The dip of the strata should preferably be gently upstream so that the resultant thrust from a gravity type-dam will be normal to the bedding planes (*see Figure 7.4*), and also any water movement will tend to be upstream into and under the reservoir bed. Alternatively, the dip might best be at right angles to the line of the valley, especially so in the case of rocks with cleavage planes.

7.6. SPECIAL FEATURES OF DAM CONSTRUCTION

Water Percolation Beneath Dam

The rock beneath a dam must, if possible, be completely impervious or it may allow either downstream leakage and thence upward hydrostatic pressure on the base of the structure to occur, or cause internal erosion of the strata underneath. Along the length of the dam, a grout curtain or cut-off trench of impervious material (*see* Warren, 1935) built into the dam foundation near the back face and carried to a suitable depth (*Figure 7.7*) may be

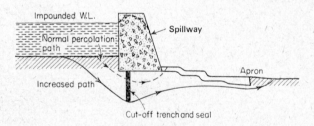

Figure 7.7. Reduction of percolation beneath a dam

necessary, or a sheet piling seal can be used; open rock joints or permeable bedding planes must be sealed by grouting; otherwise, special drainage facilities should be introduced to carry the percolating water upwards and out of sumps on the downstream face of the dam. These particular features apply equally well to any type of dam.

Adequate Spillways

Spillways must be provided on or alongside a dam to deal with flood-water overflow and the scouring action of the flood waters allowed for in the design; this should cater for the largest possible quantity of overflow (as deduced from rainfall records, etc.) and also for the effects of scour by construction of a concrete apron at the toe; this apron also allows an increased percolation path beneath the dam, thus reducing the local hydraulic gradient and pressure of any hydrostatic uplift. That sufficient outlet for the retained water is allowed during flood periods is a most important consideration, safeguarding against removal of rock from the sides of a dam and a dangerous weakening of its structure.*

Use of local stone—For a gravity dam construction, local stone effects considerable economies and should be borne in mind when choosing a suitable dam site. The main factors governing the selection of suitable stones are cost, colour and durability.

The economic cost depends mainly on:

(1) Availability—access to source, distance of transportation to site, quantity and quality of the stone;

(2) Workability of the rock strata for ease in extracting, quarrying and dressing blocks as building stones, or for crushing rock to be used as aggregate;

(3) Location of suitable quarries.

A light colour is usually preferred for masonry and thus the more durable types of igneous rocks are widely used whenever possible, and also because, generally, larger blocks can be quarried. In quarrying operations, joints giving a 'freestone' are valuable but a too close and heavy jointing in the rock will allow of easy weathering; however, some ornamental stones like marble and serpentine are certainly used, if available. The structure, texture and mineral composition of a stone are very important (Chap. 3); good abrasive resistance is needed as well as crushing and shear strength for durability, long life and a minimum of maintenance. Injurious minerals to be avoided in a stone are flint or chert, also mica in granites, gneisses, sandstones and marbles, and iron pyrites which, on exposure to the atmosphere, are subject to oxidation and change to limonite.

Summary of Requisites for a Reservoir Site Including a Gravity-type Dam†

1. A water-tight basin of ample size, with a suitable elevation.

2. A narrow outlet requiring a relatively small and economical dam possessing safe foundations, yet providing

---

* The Dolgarrog Dam failure in N. Wales (1926) was in part due to inadequate spillway provision. *See* Skeat and Hobbs, 1961, Chap. 6, page 270 et seq.

† After Bryan, K. and Lippincott, J. B., *Wat.-Supply Irrig. Pap., Wash.* No. 597A (1928) and No. 58 (1902).

3. An opportunity to build a safe and ample spillway.

4. Sufficient materials available locally from which to construct the dam, especially concrete aggregates, building stones, or rock-fill.

5. A definite assurance that the reservoir basin and catchment will not silt up in too short a time.

6. An ample and continuous water supply, both in quantity and quality, from a catchment area with a suitable local water table.

7. A practical use for the stored water or some other adequate reason to justify the cost of reservoir development.

8. A feasible distance from reservoir to consumer.

The ideal catchment area, then, is in mountain or hill country consisting of hard, compact, largely unfissured rock, where rainfall and 'run-off' is well above average (the 'run-off' in such a region can be greater than one-third of the local annual rainfall). The site chosen should have little effect on land cultivation (which is virtually non-existent), population (small) and scores particularly where land value is cheap; suitable reservoir-type valleys may be either deep and narrow or more open and glaciated. In Great Britain, the western highlands throughout Wales, England and Scotland are formed of Lower Paleozoic, Pre-Cambrian or igneous rock, which provide ample run-off characteristics, good foundation material, and suitable building stones or concrete aggregates for gravity dams, although the distance from these catchment areas to the main population centres is great.

In the Millstone Grit, Lower Carboniferous and Old Red Sandstone formations, the argillaceous rocks present can be troublesome, but the general proximity of upland areas like the Pennines to many nearby industrial towns and coalfields, makes the use of these rocks very convenient for water supply catchments.

Some references to examples of large dam schemes in Great Britain, the U.S.A., and elsewhere are given in Section 7.10 because they are instructive in the use of the principles already outlined; particularly so in the case of failures, which only too clearly illustrate the need for careful investigation and planning before a dam or reservoir construction is commenced.

7.7. OIL SUPPLIES

Oil supplies are usually obtained from a porous oil-bearing sedimentary type rock such as an open-textured sand and sandstone, or fissured limestone; here, the crude oil has accumulated along with natural gas in an underground reservoir by migrating through the rock pores in much the same way as water does, although the type of regional geological structure required for retention of an underground oil pool is precisely the reverse of that forming a suitable water aquifer in permeable strata.

7.7. OIL SUPPLIES

Oil, therefore, does not form pools and streams or underground reservoirs, in the accepted sense; because it is lighter than water, oil is usually subjected to an upward water pressure and found above its corresponding connate water level, when it sometimes seeps out at the earth's surface until perhaps the supply is exhausted. This phenomenon occurs in Trinidad, where pitch flows upwards into the famous Asphalt Lake.* However, if a suitable geological structure is present to form a seal or 'Cap rock' of impervious material, it keeps the oil permanently imprisoned underground.

The origin of oil (i.e., crude petroleum, which includes solid, liquid and gaseous hydrocarbons of the paraffin series (C_nH_{2n+2})) is a difficult question to solve, although most authorities believe an organic sedimentary origin to be practically certain. For example, consider the formation from decomposed organic matter of methane as 'marsh gas' in estuarial muds under stagnant conditions. Also at some formative intermediate stage, migration from the interstices of the oil's source rock to the point of final accumulation under the action of gravity and possibly hydrostatic head, seems normally to have taken place in an anticlinal or dome-like structure (*Figure 7.8*), with an

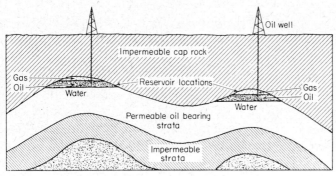

Figure 7.8. Accumulation of oil reservoirs in permeable bed folded into anticlines —diagrammatic geological section of trap structure

(Note: Oil is not found in igneous or metamorphic rock, because the conditions of their formation would decompose any pre-existing quantity of the requisite organic matter originally present in muds and oil shales)

impervious cap rock; this is commonly of clay or shale, which then creates the conditions required for artesian flow if a well is sunk into the oil reservoir. The flow may be very powerful initially, but soon a typical 'gusher' dies down and the oil has thereafter to be pumped to the surface. Frictional and capillary resistance in its pores prevent complete extraction of oil from the reservoir rock and generally only about half the total oil in storage can be extracted from wells by present-day methods of production; most important

* The Trinidad Lake Asphalt Co. supplies asphalt for road surfacing, etc., and issues literature on its activities. Oil companies (e.g. B.P. Ltd.) do likewise for the North Sea gas exploration etc.

wells are exhausted within a matter of years as the oil quantity available, unlike a supply of water, is not renewable by continuous 'recatchment'. Other typical oil traps are caused either by unconformities or faults which bury the reservoir rock under impervious strata.

The use of natural resources, e.g., coal and oil/gas for the requirements of power and modern technological processes, is rapidly consuming the earth's stored capital of energy supplied originally by the sun's effect on plant and animal life: new power sources are continually being pursued (e.g., tidal water energy) and a recent development is the growth of atomic energy power supplies.

Numerous geologists are continually employed in the task of selecting sites for oil drills throughout the world, the latest development being the exploitation of gas and oil supplies from Algeria and North Holland, and the much-publicized North Sea explorations of today.

It was the discovery of the world's largest natural gas field in Groningen, North Holland, that started the quest for oil under the North Sea. It now seems that exploitation of numerous gas fields in commercial quantities will be the end result of the vast and risky expenditure by the various oil companies involved in the race to discover the potential of the probable gas/oil bearing rock strata that underwater seismic surveys have indicated exist in this region (see Cooper (1966) and Marriott (1969).

Although 'oil' is strictly speaking a mineral resource, it has been included here because of its obvious connections with all forms of engineering construction in providing a power source for mobile plant, trucks, earth-moving equipment, etc.

7.8. GEOPHYSICAL METHODS FOR EXPLORATORY SURVEY AND SITE INVESTIGATION

Applied Geophysics as adapted for survey methods, where ground conditions favour this approach, uses the different physical properties of rocks and minerals, although only four of these properties are of real practical significance, namely those of:

1. *Local seismic wave propagation* through rocks, measured and studied in exactly the same way as shock waves are recorded by seismographs for natural earthquakes (Chap. 1), but here, for survey purposes, artificially produced by buried explosive shot charges.

2. *Electrical resistivity* variations with rock and mineral properties measured over a given distance and giving a result which is not so much a permanent characteristic of different rock types, but rather more related to their different modes of formation, grain sizes, state of consolidation, porosity and saline content.

3. *Local magnetic distortion* of the earth's overall magnetic 'field' by magnetized minerals, or rocks containing large concentrations of these minerals.

4. *Rock density* variations and their relationship to the earth's overall mean gravitational force, as used for 'gravitational' (or 'gravimetric') survey methods in oil prospecting along with Method 1.

Additionally, in earthquake regions of the earth's crust or where ground vibrations occur locally due to heavy industrial plants, dynamic study of the rock strata properties and any proposed structure to be founded upon them is usually required for the types of vibration expected, Terzaghi, 1943.* With the aid of modified 'seismic equipment', controlled vibrations are produced, sometimes mechanically, from which, at different distances from the shock source, the natural frequency and damping characteristics of the ground can be measured, together with the estimated shock effects on the vibration modes of a proposed structure.

Oil Prospecting—Use of Seismic, Electrical, Gravitational Methods

A description of geological prospecting for oil is instructive, as it embraces the use of the various methods of expert geophysical exploration—seismic, electrical, gravitational—and is backed up by more conventional drilling operations to confirm a 'strike'.

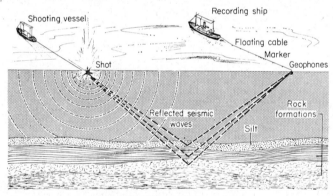

Figure 7.9. Seismic reflection technique

(Shock waves travelling from the shot are reflected by the rock layers and picked up by geophones. The geophones relay this energy to the recording ship (truck on land), where the seismograph recordings are made. Reflection shooting is more accurate, recordings are easier to interpret and it is cheaper than refraction methods)

(From *Seismic Prospecting* by courtesy of I.C.I.)

The first step is to discover and elucidate the position of reservoir structures among potential oil-bearing rock strata. As mentioned already, surface seepages, whether gaseous, liquid or solid, are often found as an indication of such oil reservoir localities and vary in nature from gases such as methane, carbon dioxide and hydrogen sulphide through light coloured water-oil

* *See also* CP 2001, No. 1, *Site Investigations* (paras. K 222, 225 and 300 etc.).

mixtures to dark solid asphaltic pitch lakes like the Trinidad example. Although these pointers do not necessarily mean that an exploitable supply is available, trial wells are usually drilled or sunk, if the local geological structure appears at all favourable for oil retention. A detailed large scale geological map of the selected area is required, with every dip carefully

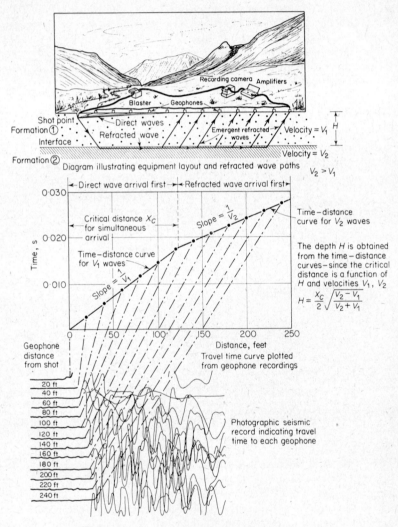

Figure 7.10. Composite illustration of seismic refraction technique

(From *Handbook of Site Investigation, etc.* by courtesy of Soil Mechanics Ltd.)

222

recorded and the exact correlation of critical strata from one locality to another needs to be undertaken.

The smallest changes (< 3 degrees of dip, say) may be vital in indicating the 'closure' and potential capacity of a possible oil reservoir and for help in choosing a high spot for the first trial drill; an expert palaeontologist may be employed also to correlate the strata by a microscopic examination and classification of their fossil content as shown in the successive rock cores obtained by the drill as it bores deeper and deeper.

The *measurement* of the resistivity and other electrical properties of the relevant strata in boreholes, and particularly for registering the positions of clays, sands and limestones, is also adopted whenever possible. If, however, the oil strata are overlain by newer sediments, perhaps unconformably, geophysical prospecting by seismic measurement at the surface may better reveal the underground rock structure.

Figure 7.11. Resistivity equipment is widely used in highway engineering to determine thicknesses of superficial deposits, and to locate construction materials such as sand and gravel

(From *Hunting Survey Review* by courtesy of Hunting Technical Services Ltd.)

Geophysical Methods, details (see Heiland (1951) and Robertshaw (1955))

1. *The seismic method* employs a type of echo sounding through rocks in which a charge is fired in a borehole from 10 to 100 or more feet deep. The reflected and/or refracted shock waves (velocities dependent upon the comparative elasticity and density of different rock strata) arrive at different

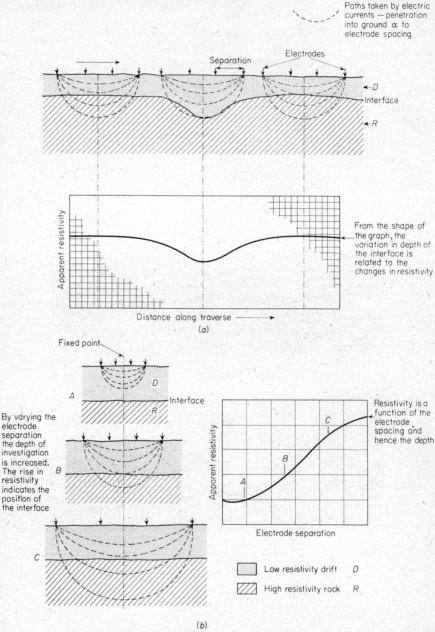

Figure 7.12. Diagrams illustrating application of resistivity method to location of buried channel and determination of depth to rock. (a) Constant electrode spacing—lateral traverse of electrode system. (b) Expanding electrode system

(Four electrodes are driven into the ground at equal distances along a straight line. A current is passed through the ground between the two outside electrodes and the potential difference between the two inner electrodes is measured to obtain the resistivity)
(From Handbook of Site Investigation, etc. by courtesy of Soil Mechanics Ltd.)

groups of carefully spaced detectors at varying times after reflection and refraction from and through the various hard rock surfaces underground. The waves are amplified in the detector 'geophones', which convert the minute ground vibrations into electrical impulses and record these photographically on moving film, which registers similarly the original explosion wave direct. The depths of the various reflecting interface rock surfaces, as calculated and correlated from a whole series of 'shothole' observations, allow the production of contoured maps showing the underground rock structure (*Figures 7.9* and *7.10*).

The seismic method has been developed for determining the relatively shallow depths of solid bedrock on most civil engineering sites and is also adaptable to mining problems; but although rapid in operation, this method is rather expensive, unsuitable for small sites, and not applicable in 'built-up' areas or for determining the depth of the water table. A prerequisite for the successful application of any seismic method is that the underlying rock formations must have a higher transmission velocity than the near-surface formations, but this is generally the case.

2. *Electrical methods*, depending in principle on the comparative measurement of the electrical resistivity and inductance of rock strata to different depths by varying the electrode separations at ground level (*Figuer 7.12b*), are extensively used in mineral surveys and are particularly suitable for shallow exploration; in oil prospecting, which requires penetration of strata to greater depths, electrical methods are rarely used except as mentioned previously in the shallowest trial bores.

Electrical methods are cheaper, quicker and more useful in determining the water table depth than other methods and are most frequently used in site surveys for civil engineering works, i.e., on large building sites, for bridges, roads, dams, tunnels, dock and harbour works, as well as for water supply surveys (*Figures 7.11* and *7.12*).

Most engineering problems needing a comprehensive site investigation can be examined by an electrical method, but confirmation of the results thus obtained by suitably positioned boreholes is desirable. Seismic, and particularly electrical methods, are the main ones used for civil engineering projects.

3. *Magnetic surveys* which are carried out on the surface (*Figure 7.13*), are necessary in igneous rock areas for locating and tracing buried rock channels in which water may accumulate and flow; for locating also where porous and permeable formations are cut by faults or igneous intrusions and to discover disused mine shafts. The methods adopted are based on the local distortion of the earth's magnetic field by minerals such as magnetite, or rocks with heavy concentrations of the iron silicate minerals, and thus provide also a useful means of detecting such mineral deposits at relatively shallow depths. Iron ores and basic igneous rocks show high magnetic

distortions, while acid igneous rocks like granites and also sediments, have a low distorting effect.

4. *In Gravitational methods*, suitable for the investigation of really large rock structures but not for most civil engineering site surveys, an extremely sensitive torsion balance (the 'Eötvös balance' or 'Gravimeter') is used to measure minute variations in gravitational force over a selected area. The probable rock mass distribution underground is interpreted from the local increase or decrease of 'g', which is dependent upon the variable density of individual rock strata. For instance, a large anticline having a denser rock core than the material of its surrounding younger earth slopes, would show a variation increase of a few milligals per mile. A milligal has value of approximately $10^{-6}g$ and this indicates both the delicacy of the instrument and the skill required in its use, again principally for oil prospecting.

To conclude—Geophysical surveyors are expert and highly experienced investigators, usually backed by the full technical resources and equipment of a specialist firm: from the engineer's viewpoint, such surveys are entirely dependent on the accurate measurement and correct interpretation of any field observations made, so that his responsibility is normally confined to the best application of the specialist's recommendations.

Figure 7.13. Operation of a Sud proton magnetometer

(The front man carries the sensing head and the U.H.F. generator mounted on a pole which is connected by a lead to the recording console carried by his partner some 30 ft behind. Continuous recordings from the head are displayed at the console and are read while the team is moving across the area to be surveyed)

(From *Hunting Survey Review* by courtesy of Hunting Technical Services Ltd.)

7.9. EXERCISES

1. (*a*) State, giving reasons, the geological features you consider should be present and what data needs to be obtained in the investigation of the site for an important dam impounding a large reservoir.

What other general information is required and what, approximately, is the relative importance of the various factors involved in choosing the position of a particular dam site?

(b) Show how the effects of glaciation can have a special influence on the siting of a dam across a suitable valley.

2. Give an account of 'resistivity surveying' with particular emphasis on its value in civil engineering operations and compare its uses with those of seismic investigations (see Figures 7.9–7.12 and CP 2001).

3. Outline the data which should be recorded in a preliminary geological survey to be carried out for establishing the line of a tunnel and investigating any problems connected with driving the heading. Note, in particular, which factors will have most influence on the engineering methods to be adopted (see Szechy, 1966).

4. A horizontal stratum of sandy soil overlies a horizontal bed of impermeable material, the surface of which is also horizontal. In order to determine the in situ permeability of the soil, a test well was driven to the bottom of the stratum and a pumping test applied. Two observation boreholes were sunk at distances of 40 and 80 ft, respectively, from the test well and water was pumped from the test well at the rate of 6·5 ft³/min until the water level became steady. The heights of water in the two boreholes were then found to be 14 and 21 ft above the impermeable bed. Find the value, expressed in feet per day, of the coefficient of permeability of the sandy soil, deriving any formula used. [From I.C.E. Pt. II examination. Apr. 1962.]

5. Describe the role of the geologist in Highway Engineering today in relation to rocks and soils, choice of route, excavation, construction methods and costs. Refer in your account to modern methods of site investigation including photogeology, geophysical prospecting (especially seismic and electrical methods) and exploratory drilling for detailed investigations at depth (see CP 2001, Site Investigation; CP 2006, Traffic Bearing Structures. Pavings; and see also Green, 1968).

6. Discuss the causes of dam and reservoir failures and indicate how these may be obviated by a correct geological approach to the engineering problems likely to be encountered on a typical dam site. (Note: See cases and references on page 231, expecially Gruner, 1963.)

7. Refer to Section 6.7, Ques. 8. Give a detailed report on the possible sites* for the dam required in this scheme and critically examine the geological conditions to be encountered along the line you recommend. Summarize your conclusions.

---

* A narrow valley with rock at no great depth is suitable for a concrete gravity, arch or buttress-type dam. For a wide valley with drift cover, an embankment-type dam of earth or rock-fill is generally more economic.

8. What are the aims of site investigation?

Describe in some detail various stages of such an investigation for a proposed office site development which will include multi-storey tall buildings. (*See* CP 2004, Section 1.2.)

7.10. FURTHER WORK—SOME SUGGESTIONS

1. Note the information available in the references for various reservoir, dam and hydro-electric projects and amplify your knowledge by referring to the originals as you think necessary. In particular, obtain a copy, if possible, of the pamphlet on 'Dams' by J. Guthrie Brown (1965).

(*a*) *Dam and Reservoir Projects*—Geology of sites, examples, *see*
  Lake Vrynwy, Claerwen Dam, N. Wales—R. C. S. Walters (1962)
  Treig Dam, Laggan Dam, Lochaber Water Power Scheme, Scotland—
  R. C. S. Walters (1962) and A. H. Naylor (1937)
  Boulder Dam, U.S.A.—C. P. Berkey (1935) and Berkey Volume, 1950
  for dams; 1951 for tunnels. (*See also* Leggett, 1962.)
  Dokan Dam, Iraq—G. M. Binnie (1959).

World Register of Dams (4 vols.) was published in 1964 by the International Commission on Large Dams (ICOLD) and contains details of almost 10,000 dams of all types in many parts of the world (as completed by the end of 1962). *Figure 7.14* shows cross-sections of two major dams recently completed and Table 7.4 lists the highest dams in the world by 1966. (Note: the most recent *ICOLD, 7th, 8th and 9th Congresses* took place in 1961, 1964 and 1967 and the 'Proceedings' and individual 'Papers' are available and worth further study.)

Table 7.4 (*a*). The 10 Highest Dams in the World
(From J. Guthrie Brown on *Dams* 1965, by courtesy of Hunting Technical Services Ltd.

Name of dam	Year of completion	Type	Height ft	Length ft	Volume of dam in cubic yards
Inguri (USSR)	U.C.	A	988	2,240	3,920,000
Nurek (USSR)	U.C.	R	984	2,395	58,860,000
Grand Dixence (Swit.)	1962	G	932	2,296	7,792,000
Vaiont (Italy)*	1961	A	858	624	460,000
Mauvoisin (Swit.)	1958	A	777	1,706	2,655,000
Contra (Swit.)	U.C.	A	754	1,206	863,000
Manicougan (Canada)	U.C.	M.A.	740	4,200	2,600,000
Bhakra (India)	U.C.	G	740	1,700	5,317,000
Oroville (USA)	U.C.	E	735	6,800	78,000,000
Glen Canyon (USA)	U.C.	A	710	1,550	4,865,000

Type of dam: A - Concrete arch; R - Rock embankment; G - Concrete gravity; M.A. - Multiple concrete arch; E - Earth embankment; U.C. - Under construction.
  * Note: Reservoir failure in 1963.

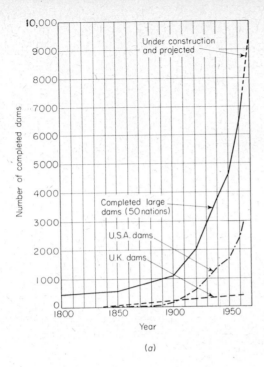

(a)

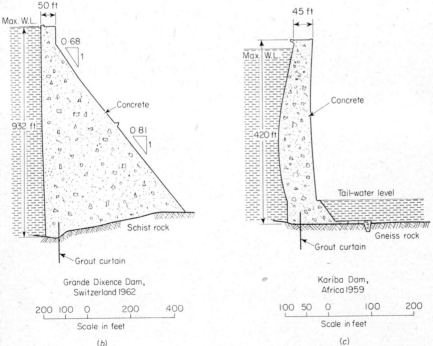

(b)

(c)

Figure 7.14. *Typical cross-section of two large modern dams: (a) Growth in construction of large dams 1800-1962. (b) Highest gravity dam in the world. (c) Slender double-curvature arch dam*

(From *Hunting Group Review* by courtesy of Hunting Technical Services Ltd. and J. Guthrie Brown)

Table 7.4 (*b*). Ten of the World's Greatest Embankment Dams based on Volume of Materials
(From J. Guthrie-Brown on *Dams*, 1965, by courtesy of Hunting Technical Services Ltd.)

Name of dam	Year of completion	Height of dam ft	Length of dam ft	Volume of materials in dam cubic yards
Fort Peck (USA)	1940	250	21,026	125,600,000
Oahe (USA)	U.C.	245	9,300	92,000,000
Mangla (Pakistan)	U.C.	380	11,000	78,000,000
Oroville (USA)	U.C.	735	6,800	78,000,000
Garrison (USA)	1960	210	11,300	66,500,000
Nurek (USSR)	U.C.	984	2,395	58,860,000
Gorky (USSR)	1955	105	40,520	57,550,000
High Aswan (UAR)	U.C.	360	16,400	55,000,000
Fort Randall (USA)	1956	165	10,700	50,200,000
Kuibyshev (USSR)	1955	148	12,405	44,298,000

The *Design and Constructional Features of Hydro-electric Dams built in Scotland since 1945*—Fulton and Dickerson, 1964, deals with the design and construction of over 50 dams in the north of Scotland.

Of the largest projects in recent years, special reference should be made to the Kariba Concrete Arch Dam (1959) on the Zambesi River (*Figure 7.14*), the High Aswan Dam on the Nile, the Roseiras Dam on the Blue Nile—all in Africa—and the Mangla Dam in Pakistan. The last three are embankment type dams under construction in 1965 (*see* Knill and Jones, 1965, and also note the Rock Mechanic references listed therein; also Binnie, 1967, for the Mangla Dam project).

(*b*) Reservoirs and Leakage Problems

Serious water leakage through permeable strata with locally impermeable rocks and regionally deep water table: Examples:

No. 3 Reservoir, Carlsbad, U.S.A. (*see* U.S. *Wat. Supply Irrig. Pap. Wash.* No. 580, 1926.

Hondo Reservoir, U.S.A. (*see* U.S. *Wat. Supply Irrig. Pap. Wash.* No. 597A, 1928).

Cedar River Dam/Reservoir, Washington, U.S.A. (*see* U.S. *Wat. Supply Irrig. Pap. Wash.* No. 597A).

(*c*) Dam Failures

These failures are of two types: (*i*) due to inadequate spillway and hence floodwater overtops the dam, (*ii*) due to a defective foundation condition or landslides.

7.10. FURTHER WORK—SOME SUGGESTIONS

Examples:

Dolgarrog Reservoir Dams, N. Wales—Last serious British failure which led to the Reservoirs (Safety Provisions) Act of 1930.*

Woodhead Dam, Sheffield, 1850.

St. Francis' Dam, Los Angeles, U.S.A., 1928.

Austin Dam, Texas, U.S.A., 1900 (*see U.S. Geol. Survey, Water Supply Papers* No. 40, 1900).

Frejus Dam, Malpasset, France, December 1959—due to faulty foundations.

Baldwin Hills Reservoir Dam, Los Angeles, U.S.A., 1963—due to faulty foundations.

Vaiont Reservoir† and *Dam* disaster, N. Italy, 1963—a landslide failure (*see* Muller, 1964, especially *Proc. Instn civ. Engrs* **28**, 1964).

The reasons for the failure of dams are supposed to be due, in the main, to geological causes and leakage problems (*see* Walters, 1964, for many examples of failures, including recent cases); *see also* Stenger, 1964, Gruner, 1963 and Leggett, 1962.

2. *Note on Rock Mechanics*

(*a*) The science of 'Rock Mechanics' has grown rapidly in recent years as a specialist service to civil engineers, for the measurement of the engineering properties of rocks in the field and in the laboratory; as rocks are required to take greatly increased loads and stresses from dams, large diameter piles under tall buildings and anchorages for prestressing systems, etc., so the critical feature of many structural designs is in the natural foundation material rather than the structure itself. After *Soil Mechanics, Rock Mechanics* is the latest development in the subject of 'strength of materials' and is being applied to assess the impact of large engineering works on the geology of sites and for assessing the intact mass strength of the natural supporting rock in particular (*see* pp. 237-238). Typical comments on one important aspect of modern dam design, as an indication of the difficult nature of many analyses, are as follows:

Stability and Anchoring of Rock Slopes

For rock stability, more than the intact rock strength alone is involved; discontinuities, such as laminations and bedding planes, contraction joints,

---

* Since this failure, the I.C.E. interim report of 1933 on Floods in Relation to Reservoir Practice (known as the *Floods Committee Report*) has provided a basis for adequate spillway design. It was revised in a 1960 Appendix with new data added up to 1957.

† This may be the worst case of mass loss of life (3,000 dead) by water in the Western Hemisphere, although when the Frejus Dam in S. France collapsed in 1959, 421 people died, and in 1960, 145 were killed in the U.S.S.R. by a similar dam failure at Kiev. The breach caused in the walls of the Baldwin Hills Reservoir, U.S.A. in December 1963 is typical of an earth dam failure of major proportions, due to defective foundations. These recent disasters are a further warning to engineers of the importance of a detailed geological study of the site and foundation of a dam, as well as the valley slopes surrounding its reservoir.

tectonic fractures, etc., and their arrangement in relation to the structure which a rock mass is to support, matter more. Such discontinuities may be clean or filled with igneous matter and hydrothermal deposits, with *in situ* weathered products of their parent rock or soils penetrated down from the surface and the behaviour of a rock mass is dependent on the mechanical properties of these fillings. Analysis is also complicated by residual internal stresses of tectonic origin, the magnitude and direction of which need to be evaluated at the design stage for their influence on the future engineering structure and its foundations. A further point of no small import for the subsequent behaviour of a rock mass, is the effect of blasting which brings a consideration of construction techniques into the picture and water seepage developing later through rock joints, etc., may affect slide conditions also.

(*b*) Obtain for personal study, a reference copy of the *Geotechnical Pamphlet No. 10, Rock Mechanics** and note especially the list given at the end thereof of recent important engineering works requiring specialist services; this report indicates the wide variety of scientific problems related to rocks, which engage the attention of practising civil engineers today. The pamphlet should form a springboard for further study—one particular topic is suggested as follows:

(*c*) As a prominent aspect of civil engineering operations requiring rocks to carry ever-increasing loads and stresses and hence need 'rock mechanics' investigation, we can cite the present scale of dam construction.

With the increasing size and weight of dams (Table 7.4) and the necessity, since the best sites are already developed, to use less favourable sites in industrialized countries, while demand for water increases, the geological problems, involving shattered, weathered, folded and jointed strata, have become much more complex. The same general axioms apply to buildings, motorways, airfields, tunnels and mines, and indeed most large-scale construction work.

3. Examine the geological aspects of the M.6 Motorway construction in Staffordshire, England (*see* Glossop, 1968, Skempton, 1964 and 1965),† with special reference to the many obstacles encountered in the glacial valley known as Walton's Wood—a particularly interesting example of road construction difficulties—(*see also* Rodin, 1964, p. 34 et seq.).

4. Review the recent Vaiont Reservoir disaster‡ (N. Italy, October, 1963) and draw what conclusions you can about the great strength of modern dams when built on a secure foundation. Consider the dynamics of the reservoir region and the changes that could cause huge landslides.

---

* Issued by Rocks Mechanics Ltd., London, S.W.3.
† *See The Guardian*, London, Nov. 15th, 1963.
‡ *See* references cited on pages 211, 213 and 236.

(5) (*a*). Assemble what information you can on the newly-opened Mount Blanc Road Tunnel* and also on the much earlier Simplon Tunnel through the European Alps and compare these projects with the Mersey Road Tunnel in England, starting from Fox, 1907, and Boswell, 1937. (Similarly compare the Dartford Tunnel, Kell (1963), and the Tyne Tunnel, Prosser (1968).)

(*b*) *Figure 7.15* shows the McAlpine Tunnelling Machine developed for soft tunnelling and used in boring the Toronto subway, Canada, and London Underground Tube railways (e.g., for the Victoria Line tube construction at present). Further information is available from London Transport Publicity and Sir Robert McAlpine and Sons Ltd., London, W.1.†

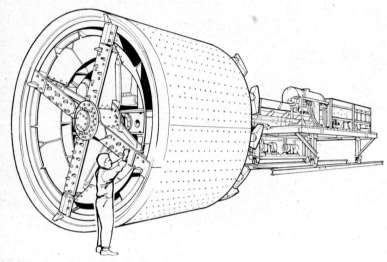

Figure 7.15 The McAlpine tunnelling machine—a mechanical drum digger shield for soft rock drift

(On the working face, the arms across the front of the shield rotate and the cutting teeth dislodge the clay in front of them, which is then fed by the paddles through the central hole on to the conveyor belt to be carried away to the rear and soil removal equipment tubes, etc.
Four of these all-hydraulic machines, designed and built by Sir Robert McAlpine and Sons, were used for work on the Victoria line extensions in the London clay)
(By courtesy of Sir Robert McAlpine and Sons)

(*c*) Find out all you can about the Severn Railway Tunnel (1886) and cementation thereof during 1929–30;‡ collect also information on the Channel Tunnel (*see* Bruckshaw *et al.* 1961), which project is now in course of revival and the subject of recent intensive geological survey.§

* *See The Guardian*, London, 16th July, 1955 and Mayer, 1963.
† *See also The Guardian*, London, 28th July, 1965, and Hay, *et al.* 1965.
‡ *See* Carpmael, 1931–32.
§ *See The Guardian*, London, 11th June, 1965.

EPILOGUE

In concluding this book, one final note.

All engineers, young and older, experienced and inexperienced, cannot do better than read Ralph Peck's paper* on the role of active and intensive individual experience in developing personally throughout one's whole professional life the twin *arts* of engineering geology and soil mechanics in the practice of *geotechnology*, a continuous and rewarding process.

REFERENCES AND BIBLIOGRAPHY

Berkey,† C. P. (1935). 'Geology of Boulder and Norris Dam Sites'. *Civ. Engng* **5**. New York.

Binnie,‡ G. M. (1959). 'The Dokan Project'. *Proc. Instn civ. Engrs* **14**, Oct. Paper 6389–90.

Binnie, G. M. (1967). 'The Mangla Dam'. *Proc. Instn civ. Engrs* **38**, Nov.

Bishop, A. W. (1969). 'The Aberfan Disaster: Technical Aspects.' *Proc. Instn civ. Engrs* **42**, Feb.

Boswell, P. G. H. (1937). 'The Geology of the New Mersey Tunnel'. *Proc. Lpool geol. Soc.* **17**.

Bruckshaw, J. M. (1961). 'The Work of the Channel Study Group 1958–60'. *Proc. Instn civ. Engrs* **18**, 149 Paper 6509.

Carpmael, P. (1931–32). 'Cementation of the Severn Railway Tunnel'. *Proc. Instn civ. Engrs* **234**.

Chapman, S. C. (1923). 'Notes on Concrete Exposed to a Moorland Water'. *Trans. Instn Wat. Engrs* **28**, 25.

Cooling, L. F. (1962). 'Field Measurements in Soil Mechanics'. *Geotechnique* **12**, 89.

Fox, F. (1907). 'The Simplon Tunnel'. *Proc. Instn civ. Engrs* **168**.

Fulton, A. A. (1952–3). 'Civil Engineering Aspects of Hydroelectric Developments in Scotland'. *Proc. Instn civ. Engrs* **1**, 248.

Fulton, A. A. and Dickerson, L. H. (1964). 'The Design and Construction Features of Hydro-electric Dams Built in Scotland since 1945'. *Proc. Instn civ. Engrs* **29**, Dec.

Green, P. A. (1968). '*Informal Discussion 21232/94, 12 March 1968 on Ground and Materials Investigations for Road Schemes—Needs and Methods*'. London; I.C.E.

Gruner, E. (1963). 'Dam Disasters'. *Proc. Instn civ. Engrs* **24**, Paper 6648.

Guthrie-Brown, J. (1965). 'Dams'. *Hunting Gp. Rev. No. 1.*

Halcrow, W. T., Brook, G. B. and Preston, R. (1928). 'The Corrosive Attack of Moorland Water on Concrete'. *Trans. Instn Wat. Engrs* **33**, 187.

Harding, H. J. B. (1949). 'Site Investigations Including Boring and other Methods of Sub-surface Exploration'. *Proc. Instn civ. Engrs* **32**, April.

Hay, J. D., Hughes, H. M. and Wrathall, R. D. (1965). 'The Bretby Tunnelling Machine'. *Proc. Instn civ. Engrs* **30**, 649. Paper 6830.

Heiland, C. A. (1951). *Geophysical Exploration*. New York; Prentice Hall.

Henkel, D. J. (1957). 'Investigation of two Long-term Failures in London Clay Slopes at Wood Green and Northolt'. *Proc. Int. Conf. Soil Mech.* **2**, 315.

* *See* Peck, 1962, for a masterly review of *Geotechnology* and reference list to date, *see* Glossop, 1968.

† *See also* Berkey volume, Geological Society of America. Application of Geology to Engineering Practice: 1950 for Dams, 1951 for Tunnels.

‡ Describes the design and construction of a concrete arch dam in the Dokan Gorge, Iraq, and is especially instructive in the methods for, and use of, a grouted cut-off curtain.

REFERENCES AND BIBLIOGRAPHY

Hepworth, J. V. (1967). 'Photogeological Recognition of Ancient Orogenic Belts in Africa'. *Proc. geol. Soc.* No. 1641, 166.

Horslev, M. J. (1949). *Sub-surface Exploration and Sampling of Soils for Civil Engineering Purposes.* Am. Soc. civ. Engrs Soil Mechanics and Foundations Division Waterways Exp. Stn. Vicksburg, Mississippi.

Hutchinson, J. N. (1968). 'Stability of Slopes in Fissured Clays'. *Informal Discussion.* British Geotechnical Society, 17 Jan., London; I.C.E.

Hutchinson, J. N. (1969). 'Coastal Landslides at Folkestone Warren, Kent.' *Geotechnique* **19**, No. 1, 6.

Ischy, E. and Glossop, R. (1962). 'An Introduction to Alluvial Grouting'. *Proc. Instn civ. Engrs* **21**, 449. Paper 6598; 'Discussion': *Proc. Instn civ. Engrs* **23**, 205.

Kell, J. (1963). 'The Dartford Tunnel'. *Proc. Instn civ. Engrs* **24**, 359.

*Knill, J. L. and Jones, K. S. (1965). 'The Recording and Interpretation of Geological Conditions in the Foundations of the Roseiras, Kariba and Latiyan Dams'. *Geotechnique* **15**, No. 1, 94.

Langfors, U. and Kohlstrom, B. (1964). *Rock Blasting.* New York; Wiley.

McIldowie, G. (1934). 'The Construction of the Silent Valley Reservoir, Belfast Water Supply'. *Mins Proc. Instn civ. Engrs* **239**, 465.

Mott, P. G. (1963). 'Aerial Methods of Surveying for Civil Engineering'. *Proc. Instn civ. Engrs.* **26**, 497. Paper 6728.

Muller, L. (1964). 'The Rock Slide in the Vaiont Valley'. *Rock Mechanics and Engineering Geology.* **II**, 3–4, 148.

Muskat, M. (1937). *The Flow of Homogeneous Fluids through Porous Media.* New York; McGraw-Hill.

Naylor, A. H. (1937). 'The Second Stage Development of the Lochaber Water Power Scheme'. *J. Instn civ. Engrs* **5**, 3.

Nelson, A. and Nelson, K. D. (1967). *Dictionary of Applied Geology, Mining and Civil Engineering.* London; Newnes.

Newberry, J. (1968). 'Influence of Geology on the Design of Dams.' ICOLD *Informal Discussion.* 17th June, London; I.C.E.

Norman, J. W. (1967). Synopsis of paper given at a Geological Society of London Engineering Group Meeting at Cardiff, 28 Sept. (*see* page 166)

Obert, L. and Duval, W. I. (1967). *Rock Mechanics and the Design of Structure in Rock.* New York; Wiley.

Penman, H. L. (1956). 'A Field Piezometer Apparatus'. *Geotechnique* **6**, No. 2, 57 and 137.

Penman, H. L. (1961). ' A Study of the Response Time of Various Types of Piezometers'. In: *Pore Pressure and Suction in Soils.* London; Butterworths.

Prosser, J. R. and St. C. Grant, P. A. (1968). 'The Tyne Tunnel: Planning of the Scheme'. *Proc. Instn civ. Engrs* **39**, Paper 7068.

Reynolds, H. R. and Protopadakis, P. (1946). 'Stability of Earth Slopes'. *Proc. Instn civ. Engrs* **41**, No. 479, 178.

Rodin, S. (1964). 'Earthworks'. *Muck Shifter and Bulk Handler* **22**, No. 4. April.

Scott, K. F., Reeve, W. T. N. and Germond, J. P. (1968). 'Farahnaz Pahlavi Dam at Latiyan'. *Proc. Instn civ. Engrs* **39**, 353, Paper 7073.

Skeat, W. O. and Hobbs, A. T. (1961). Editors. *Manual of British Water Engineering Practice.†* 2nd edn. London; Institution of Water Engineers.

Skempton, A. W. (1946). *Earth Pressures and the Stability of Slopes:* In: *The Principles and Applications of Soil Mechanics.* (A record of four lectures: Cooling, L. F., Glossop, R., Markwick, A. M. D.). London; Institution of Civil Engineers.

* *See* Scott, *et al.*, 1968.
† A standard reference work and particularly good on legal aspects of water engineering.

Skempton, A. W. (1948). 'The Rate of Softening in Stiff Fissured Clays, with Special Reference to London Clays'. *Proc. 2nd Int. Conf. Soil Mech.* **2**, 50.

Skempton, A. W. (1964). 'The Long-term Stability of Clay Slopes'. *Geotechnique* **14**, 77: (1965) 6th Int. Conf. S.M.F.E., 235, **3**, 278–280.

Skempton, A. W. and Henkel, D. J. (1957). 'London Clay in Deep Borings at Paddington, and the South Bank'. *4th Int. Conf. Soil Mech. and Foundation Engng.*

Skempton, A. W. and Henkel, D. J. (1961). 'Pore Pressures in London Clay'. In: *Pore Pressure and Suction and Soils.* London; Butterworths.

Stenger, F. (1964) (Compiler). *Bibliography of Dam Failures.* Denver, Colorado; Bureau of Reclamation.

Széchy, Károly (1966). *The Art of Tunnelling.* Budapest; Akadémiai Kiadó.

Terzaghi, K. (1943). *Theoretical Soil Mechanics.* New York; Wiley.

Terzaghi, K. (1960). *From Theory to Practice in Soil Mechanics.* New York; Wiley.

Terzaghi, K. and Peck, R. V. (1948). *Soil Mechanics in Engineering Practice.* New York; Wiley.

Toms, A. H.* (1953). 'Recent Research into the Coastal Landslides at Folkestone Warren, Kent, England'. *Proc. 3rd Int. Conf. Soil Mech.* **2**, 288.

Walters, R. C. S. (1962). *Dam Geology.* London; Butterworths.

Walters, R. C. S. (1964). Also 'The Impact of Geology on Dams and Reservoirs'. *Muck Shifter and Bulk Handler* **22**, No. 4. April.

Whittle, G. (1967). 'Photogeology in Overseas Geological Aid and Aspects of the Work of the Photogeological Division of the Institute of the Geological Sciences.' *Proc. geol. Soc.* No. 1641, 163.

Willox, W. A. (1965). 'Photointerpretation in Geological and Soils Mapping for a Major Road Project in Spain'. *Proc. goel. Soc.* No. 1629, 5.

Winter, T. S. R. (1950–51). 'The Silting of Impounding Reservoirs'. *J. Inst. civ. Engrs* **35**, 65.

Various authors (1964). 'Vaiont Reservoir and Dam Failure N. Italy, 1963'. *Proc. Instn civ. Engrs* **28**.

SELECT BIBLIOGRAPHY

Briggs, H. (1929). *Mining Subsidence.* London; E. Arnold.

Brunton, D. W. and Davis, J. A. (1922). *Modern Tunnelling.* New York; Wiley.

Capper, P. L. and Cassie, W. F. (1963). *The Mechanics of Engineering Soils.* 4th rev. edn. London; Spon.

Capper, P. L. and Cassie, W. F. (1967). *Problems in Soil Mechanics.* London; Spon.

Casagrande, A. (1961). 'Control of Seepage through Foundations and Abutments of Dams'. *Geotechnique* **11**, No. 3.

Cooper, B. and Gaskell, T. F. (1966). *North Sea Oil—The Great Gamble.* London; Heinemann.

Goodman, L. J. and Karol, R. H. (1968). *Theory and Practice of Foundation Engineering.* London: Collier–Macmillan.

Harza, L. F. (1935). 'Uplift and Seepage under Dams in Sands'. *Trans. Am. Soc. civ. Engrs* **100**, 1352.

Lane, E. W. (1935). 'Security from Under Seepage'. *Trans. Am. Soc. civ. Engrs* **100**, 1235.

Lane, R. G. (1964). 'Rock Foundations'. *Proc. 8th Int. Conf. Large Dams.* Edinburgh; ICOLD.

* *See* Hutchinson (1969).

SELECT BIBLIOGRAPHY

Lapworth, H. (1911). 'The Geology of Dam Trenches'. *Trans. Ass. Wat. Engrs* **16**.

Leggett, R. H. (1962). *Geology and Engineering*. 2nd Edn. New York; McGraw-Hill.

Leonards, G. A. (1962). *Foundation Engineering*. New York; McGraw-Hill.

Little, A. L. and Vail, A. J. (1961). 'Some Developments in the Measurement of Pore Pressure' in *Pore Pressure and Suction in Soils*. London; Butterworths.

Marriott, G. B. and Sutton, V. J. R. (1969). 'Production of North Sea Gas.' *Proc. Instn civ. Engrs* **42**, 439.

Minikin, R. R. (1948). *Structural Foundations*. London; Crosby Lockwood.

Naylor, A. H. (1936–7). 'Lochaber Water Power Scheme'. *J. Instn civ. Engrs* **5**, No. 4.

Power, G. L. (1950). 'The Geophysical Investigation of Underground Water Supplies'. *J. Instn Wat. Engrs* **4**, 237.

Ries, H. and Watson, T. L. (1937). *Engineering Geology*. 2nd edn. New York; Wiley.

Robertshaw, J. and Brown, P. D. (1955). 'Geophysical Methods of Exploration and their Application to Civil Engineering Problems'. *Proc. Instn civ. Engrs* **4**, Pt. 1, 644.

Sonbern, J. F. (1951). *Engineering Geology in the Design and Construction of Tunnels*. Geological Society of America.

Taylor, D. W. (1948). *Fundamentals of Soil Mechanics*. New York; Wiley.

Terzaghi, K. (1943). *Theoretical Soil Mechanics*. New York; Wiley.

Tschebotarieff, G. P. (1951). *Soil Mechanics Foundations and Earth Structures*. New York; McGraw-Hill.

Walters, R. C. S. (1949). 'Some Geophysical Experiences in Water Supply'. *J. Instn Wat. Engrs* **3**, 436.

Warren, J. R. (1935). 'Cut-off Trenches in Relation to Test Bores'. *Wat. and Wat. Engng* **37**, No. 449, 291.

War Office (1945). 'The Location of Underground Water by Geological and Geophysical Methods'. *Military Engineering*. Vol. 6. Water Supply Suppl. No. 1. London; H.M.S.O.

Various authors (1963). *Grouts and Drilling Muds in Engineering Practice*. The British National Society of the International Society of Soil Mechanics and Foundation Engineering. London; Butterworths.

FURTHER READING ON ROCK MECHANICS

A.S.T.M. (1967). 'Determination of Stress in Rocks'. *Spec. tech Publ. No. 429*. Philadelphia, U.S.A.

Banks, J. A. (1957). 'Allt-na-Lairige Prestressed Concrete Dam'. *Proc. Instn civ. Engrs* **6**, 409.

Bishop, A. W. (1955). 'The Use of the Slip Circle in the Stability Analysis of Slopes'. *Geotechnique* **5**, 7.

Bruckshaw, J. M. and Dixy, F. (1934). 'Groundwater Investigation by Geophysical Methods'. *Wat. and Wat. Engng* **36**, (433) 261; (435) 368.

Capper, P. L. and Cassie, W. F. (1967). *Problems in Soil Mechanics*. London; Spon.

Cedergren, H. R. (1967). *Seepage, Drainage and Flow Nets*. New York; Wiley.

C.I.R.I.A. Research Bulletins* 1–3 (1964–5–6) et seq., for example, *see* Monar Dam, Ref. No. 34; Stithians Dam, Ref. Nos. 11, 13, 24, 25, 26, 30, 38, 39; Clywedog Dam, 49.

Cooling, L. F. (1941–42). 'Soil Mechanics and Site Exploration'. *J. Instn civ. Engrs* **18**, 47.

* Lists recent research reports on Rock/Soil mechanics and foundations of Dams, etc., including a new *Bibliography of Rock Mechanics* (Lancaster-Jones, 1966).

Creager, W. P., Hinds, J. and Justin, J. D. (1945). *Engineering for Dams* (3 vols.). New York; Wiley.

Fitt, R. L., Marwick, R., Whitaker, F. W. A. and Corney, J. V. (1967). 'The Roseiras Dam, Sudan'. *Proc. Instn civ. Engrs* **38**, 21, 53, Papers 7047 and 7048.

Fitt, R. L., Marwick, R., Whitaker, F. W. A. and Corney, J. V. (1968). 'Discussion'. *Proc. Instn civ. Engrs* **40**, 83.

Glossop, R. (1968). 'The Rise of Geotechnology and its Influence on Engineering Practice.' *Géotechnique* **18,** No. 2, 107.

Hammond, Rolt (1967). *Modern Foundation Methods*. London; McClaren.

Japanese National Committee on Large Dams (1967). *Earthquake Resisting Design Features of Dams in Japan*. The Committee.

John, K. W. (1967). 'An Approach to Rock Mechanics'. *Proc. Am. Soc. civ. Engrs* **88**, SM: 1, 1–30.

Krynine, P. D. and Judd. J. W. (1957). *Principles of Engineering Geology and Geotechnics*. New York; McGraw-Hill.

Lancaster-Jones, P. F. F. (1966). *Bibliography of Rock Mechanics*. London; The Cementation Co., Ltd.

Little, A. L. and Price, V. E. (1958). 'The Use of an Electric Computer for Slope Stability Analysis'. *Geotechnique* **8**, 113.

Mayer, A. (1963). 'Recent Work in Rock Mechanics'. *Geotechnique* **13**, No. 2, 97.

Morgan, H. D., Scott, P. A., Walton, R. J. C. and Faulkner, R. H. (1953). 'The Claerwen Dam'. *Proc. Instn civ. Engrs* **2**, 249.

Paton, J. (1956). 'The Glen Shira Hydro-electric Project'. *Proc. Instn civ. Engrs* **5**, No. 5, 593.

Price, D. G. and Knill, J. L. (1967). 'The Engineering Geology of Edinburgh Castle Rock'. *Geotechnique* **17**, 411.

Price, D. G. *et al.* (1967). 'Foundations of Multi-story Blocks in the Coal Measures with Special Reference to Old Mine Workings'. *Proc. Geol. Soc.* 1637, 3.

Price, D. G. *et al.* (1967). 'Problems of Shallow Mine Workings'. *Proc. Geol. Soc.* 1637, 8.

Peck, R. B. (1962). 'Art and Science in Sub-surface Engineering'. *Geotechnique* **12**, No. 1, March.

Reynolds, H. R. (1961). *Rock Mechanics*. London; Crosby Lockwood.

Sabarly, F. (1968). 'Les injections et les drainages de fondation de barrages.' *Geotechnique* **18,** No. 2.

Semenya, Carol (1952). 'The Most Recent Dams by the S.A.D.E. in the Eastern Alps'. *Proc. Instr. civ. Engrs* **1**, No. 5, 508.

Sharpe, C. F. S. (1938). *Landslides and Related Phenomena*. New York: Columbia Univ. Press.

Shepard, E. R. (1936). 'The Application of Geophysical Methods to Grading and other Highway Construction Problems'. *Proc. Highw. Res. Bd.* **16**, 282.

Sherard, J. L. *et al.* (1963). *Earth and Earth-rock Dams*. New York; Wiley.

Stagg, K. G. and Zienkiewicz (Eds.) (1968). *Rock Mechanics in Engineering Practice*. London: Wiley.

Talobre, J. (1957). *La Mechanique des Rockes*. Paris; Dunod.

Terzaghi, K. (1962). 'Measurement of Stresses in Rocks'. *Geotechnique* **12**, No. 2.

Various authors (1965). 'Symp. on Rock Mechanics'. *Proc. Geol. Soc.* No. 1629, Mar. 1966.

Various authors (1967). 'Symp. on Rock Slopes'. *Proc. Geol. Soc.* No. 1641, 154.

Various authors (1967). Engng Group Meeting. Cardiff, S. Wales: (1968) *Proc. Geol. Soc.* No. 1647, Aug.

Various authors (1967). Symp. on Computer Applications in the Earth Sciences. *Proc. Geol. Soc.* No. 1642, 183.

SELECT BIBLIOGRAPHY

Various authors (1969) March. 'Coastal Problems.' Engng Gp Meeting. London. *Geol. Soc. Circ.* 150, Feb. p. 3.

Conference Proceedings etc.

'The International Society of Rock Mechanics' (*see* for example, *Proc. 1st Int. Congress*, Lisbon, 1966*) and also of 'Soil Mechanics and Foundation Engineering' (*6th Int. Conf.* held at Montreal, 1965), publish relevant papers and articles on developments in these sciences via the Proceedings of International/National Congresses or Conferences and through related British National Societies and Committees (*see also Proceedings of the Geotechnical Conference*, Oslo, 1967, *Proceedings of Conference on Earth Pressure Problems*, Brussels, 1959 and '*In situ* Investigations in Soils and Rocks Conf.' London. May 1968, I.C.E.).

Individual reports appear in *Geotechnique* (for foundations, embankments, earth dams, etc., especially), the international journal published by the Institution of Civil Engineers, London, under whose auspices the British Geotechnical Society meets.

Papers on specific dam projects appear from time to time in the *Proceedings of the Institution of the Civil Engineers*, and *Geotechnique*, and the London Geological Society (*Engineering Group Quarterly Journal* commenced publication as Volume 1, November 1, 1967) also publishes very useful papers on geological aspects of engineering projects, such as foundation problems for buildings, tunnels, dams, as well as general rock mechanics research data and advice. (*International Journal of Rock Mechanics and Mining Sciences* commenced publication with Volume 1, Number 1, 1963.)

The R.R.L. and B.R.S. also publish some technical papers mainly formation, drainage (embankments, cuttings, etc.) in roads and foundations engineering, respectively, in relation to Soil Mechanics—soil problems and sometimes specific geological topics (for example, R.R.L. Ref. Nos. 39, 40, 47, LR 85 and 116—Studies of Keuper Marl, R.R.L. Research Report No. 85 (1966), as described in (1968) *Proc. Geol. Soc.* No. 1647. Aug.

* A review of present knowledge in Rock Mechanics. *See also Int. Symp.* Madrid, 1968.

INDEX

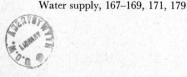